D1329602

Molecular Magnetism and Magnetic Resonance Spectroscopy

FUNDAMENTAL TOPICS IN PHYSICAL CHEMISTRY

Editor: **Harold S. Johnston,** *University of California, Berkeley*

HANNAY	*Solid-state Chemistry*
MYERS	*Molecular Magnetism and Magnetic Resonance Spectroscopy*
SOMORJAI	*Principles of Surface Chemistry*
STRAUSS	*Quantum Mechanics: An Introduction*
WESTON and SCHWARZ	*Chemical Kinetics*

ROLLIE J. MYERS
University of California, Berkeley

Molecular Magnetism and Magnetic Resonance Spectroscopy

PRENTICE-HALL INC.
Englewood Cliffs, New Jersey

FUNDAMENTAL TOPICS IN PHYSICAL CHEMISTRY
Editor: **Harold S. Johnston**

Printed in the United States of America
Library of Congress Catalog Card No. 72-5319
ISBN: 0-13-599696-1

10
9
8
7
6
5
4
3
2
1

PRENTICE-HALL INTERNATIONAL, INC., *London*
PRENTICE-HALL OF AUSTRALIA, PTY. LTD., *Sydney*
PRENTICE-HALL OF CANADA, LTD., *Toronto*
PRENTICE-HALL OF INDIA PRIVATE LIMITED, *New Delhi*
PRENTICE-HALL OF JAPAN, INC., *Tokyo*

Contents

Foreword

THE EMPHASIS IN PHYSICAL-CHEMICAL RESEARCH has shifted over the past generation from the macroscopic view of thermodynamics, electrochemistry, and empirical kinetics to the molecular view of these and other areas. Statistical mechanics has provided a practical method for the evaluation of certain types of thermodynamic data. The microscopic viewpoint of solid-state physics has injected new intellectual vigor into electrochemistry, surface chemistry, and heterogeneous catalysis. Recent developments in molecular beams have turned chemical kinetics into one of the most active fields of physical chemistry. Thus most areas of physical chemistry now present two aspects, the old and the new; and this dichotomy is reflected in the teaching of physical chemistry.

The usual curriculum in chemistry includes a year's course in physical chemistry given during the sophomore or junior year. In some cases, teachers of this course try to retain all the old physical chemistry and to introduce all the new topics as well; then the sheer volume of material is overwhelming. In other cases, an effort is made to cover all the material from the modern point of view, starting with quantum mechanics; this approach tends to neglect (or treat in only a superficial way) complicated molecules, solutions, and the liquid state. It ought to be recognized that physical chemistry is not a body of knowledge that must be covered but rather a set of methods for predicting chemical events. Each of the methods should be understood. It is desirable to present chemical thermodynamics by the macroscopic approach, lest this method of science be omitted from the student's education. Next, it is imperative to teach the student a small amount of rigorous, Schrödinger-type quantum mechanics, with some—but not all—applications to molecular structure and molecular spectroscopy. The student of physical chemistry must also receive an introduction to statistical mechanics but not a survey of all its applications. With a rigorous, though limited, introduction to thermodynamics, quantum mechanics, and statistical mechanics, the student can readily be introduced to the methods and viewpoints of chemical kinetics, solid-state chemistry, and other areas of physical chemistry.

The ideal textbook in physical chemistry might contain a condensed, clear exposition of each *method of prediction* in chemistry, with extra reading material and references. Such a book would be long, and a course based on it inflexible. It therefore seems desirable to have available a series of relatively brief texts that can help serve the same purpose and that can give the teacher greater flexibility in his choice of teaching materials.

This then is the aim of the series Fundamental Topics in Physical Chemistry. Each author has been urged to make his goal student insight into one basic area in physical chemistry.

Harold S. Johnston

Preface

FIRM THEORETICAL BASIS for an understanding of the magnetic properties of molecules was first established by J. H. Van Vleck in his book, *The Theory of Electric and Magnetic Susceptibilities*, which was first published in 1932. In this book Van Vleck showed how the new quantum mechanics could explain much of the old field of magnetism. Since that time, both crystal field theory and magnetic resonance spectroscopy have been added to this field, and the study of molecular magnetism has now taken on new dimensions.

This book was written with the assumption that both Van Vleck's methods and the spin Hamiltonian developed for magnetic resonance could be presented in a unified manner suitable for study by advanced undergraduates or first-year graduate students. We have had to assume that the reader has some previous knowledge of modern quantum mechanics, but we have also tried to explain some of its more important principles in the body of the text. In particular, we have tried to show

how the vector model, as adapted from the commutation rules of quantum mechanics, can serve as a bridge between the classical and quantum theories of magnetism. In recent years, it has become fashionable to reject the vector model as a vestige of an old and abandoned quantum mechanics, but we feel that it still has some usefulness—if properly used.

The reader is expected to consult other works while using this text and a number are given in the references. Molecular magnetism touches on a wide range of topics and the limitations of time and space have forced us to be very brief in some parts of the text. On the historical side, the reader may want to learn more of the role of J. H. Van Vleck. A very interesting biographical sketch has been published: P. W. Anderson, "Van Vleck and Magnetism," *Physics Today*, **21,** 23 (1968). From this record one can readily see how Van Vleck's work has influenced all aspects of the study of magnetism.

A number of old and present graduate students read over many of the chapters in this book. I would like to particularly thank James Chang, Michael Coggiola, Thomas Hynes, Akira Jindo, Richard Wilson, and Joyce Yarnell for their criticisms. Mrs. Nancy Monroe not only typed all the various drafts of the manuscript, but she also drew many of the figures used in the book. I would also like to thank Professors Hs. H. Günthard and E. B. Wilson, Jr. for their hospitality at the start and at the end, respectively, of my work on the manuscript.

Rollie J. Myers

A list of symbols used in this text:

Quantum Vector Operators

$\mathbf{S}, \mathbf{L}, \mathbf{J}, \mathbf{I},$

Quantum Operators

$\mathsf{H}, \mathsf{S}_z, \mathsf{L}_z, \mathsf{J}_z, \mathsf{I}_z,$

Classical Functions (A.4.)

$\mathscr{H}, \mathscr{L}$

Classical Vectors

$\mathbf{S}, \mathbf{L}, \mathbf{J}, \mathbf{I}, \mathbf{H}$

Quantum Numbers or Scalars

S, L, J, I, H

1

Molecular Magnetism

WHEN A MOLECULE is held in a uniform magnetic field, it never experiences a net force which would tend to make it move with or against the field. This is a result of the fact that all molecules are "magnetically neutral" and individual N and S poles do not exist. In a uniform electric field, of course, molecules may experience a net force. This is because positive and negative charges (or monopoles) do exist and ionized molecules are not electrically neutral. A molecule may experience a torque in a uniform magnetic field, for its energy can depend upon its orientation in the field. This torque arises from the interaction of the field and a molecular magnetic dipole moment. Since the most common form of magnetic dipole is the bar magnet or compass needle, it is common to think of magnetic dipoles as being made up of separated N and S poles. Such magnetic monopoles do not exist in nature; it is better to consider a magnetic dipole as arising from a current loop formed by a rotating charge. Molecules are composed of electrons moving around nuclei, so it is easy to see how molecules might possess magnetic dipole moments. The situation is complicated by the fact that electrons also possess an intrinsic magnetic dipole moment, which is independent of its ordinary orbital motion, and the total magnetic moment of the molecule is a sum of orbital and intrinsic moments.

It will take some time before we can work out this total magnetic moment for any molecule and we shall have to be content to approach this problem in easy steps. We shall first show that the magnetic moments of electrons are related to their angular momenta. A study of the magnetic properties of molecules is largely a study of angular momenta in molecules, and such a study must rely heavily upon quantum mechanics.

1.1. Magnetic Dipole Moments

In Fig. 1.1 we illustrate the fundamental concepts for magnetic dipole moments. One could assume that a magnetic dipole moment was formed from two poles ($+m$ and $-m$) which are separated by a distance d.

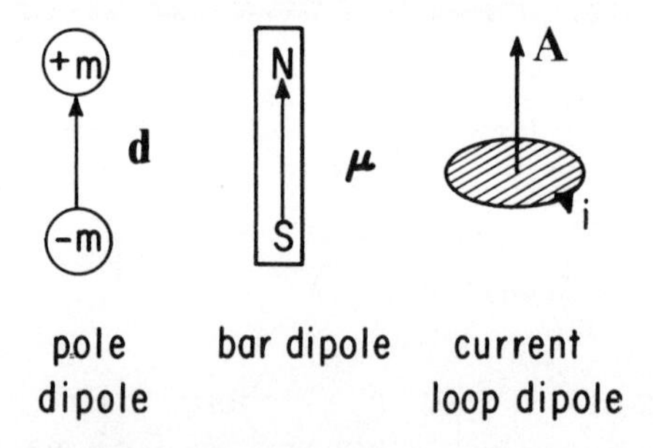

Figure 1.1. *Magnetic dipole moment.*

The dipole moment μ would be equal to the product of the magnetic monopole m times d. In vector notation this is

$$\boldsymbol{\mu} = m\mathbf{d}. \tag{1.1a}$$

Since magnetic monopoles do not exist, Eq. (1.1a) is not very useful.

The bar magnet is the popular concept of a magnetic dipole. It has a permanent moment, for the moments of individual atoms have been aligned in a strong magnetic field, and in certain materials, such as soft iron, they will stay aligned when the field is removed. One end of the bar magnet is called N and the other S. If a bar magnet is used as a compass, the N end will point north, so N means north-seeking pole. In the bar dipole the dipole moment $\boldsymbol{\mu}$ points from the S to the N end of the magnet.

The basic source of magnetic dipoles is the current loop. For a loop of current i,

$$\boldsymbol{\mu} = i\mathbf{A} \tag{1.1b}$$

where the area of the loop A forms a vector normal to the loop, as shown in Fig. 1.1. Since current is equivalent to a moving charge, one can relate it to an electrostatic charge q and its angular frequency ω. The relation is

$$i = \frac{q\omega}{2\pi c} \tag{1.2}$$

where c, the speed of light, is introduced to give magnetic units for current. Since the area of a loop is πr^2, we can rewrite Eq. (1.1b) as

$$\boldsymbol{\mu} = \frac{qr^2}{2c}\boldsymbol{\omega}. \tag{1.3}$$

The angular frequency ω follows a right-hand rule, as shown in Fig. 1.2.

The rotating charge in Fig. 1.2 possesses both a magnetic moment

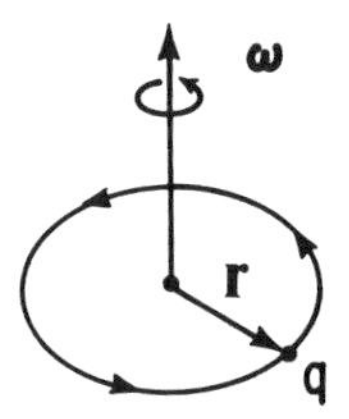

Figure 1.2. *Current loop from a rotating charge.*

$\boldsymbol{\mu}$ and an angular momentum $\mathbf{P}$. The two are proportional to each other, so

$$\boldsymbol{\mu} = \gamma\mathbf{P} \tag{1.4}$$

where γ is properly called the magnetogyric ratio. The angular momentum $\mathbf{P}$ is equal to $I\boldsymbol{\omega}$, and I, the moment of inertia of a rotating mass

m, is equal to mr^2. When combined with Eq. (1.3) these relations give

$$\gamma = \frac{q}{2mc}$$

or

$$\boldsymbol{\mu} = \frac{q}{2mc}\mathbf{P}. \tag{1.5}$$

The magnetogyric ratio depends upon the ratio of the charge and the mass of the rotating particle.

One of the results of quantum mechanics is that rotating bodies can have only certain values of angular momentum. In the hydrogen atom, for example, the electron has quantized angular momentum and it is quantized in units of $h/2\pi$ where h is Planck's constant. Since we will use $h/2\pi$ in many equations we will use the special symbol $\hbar$ for it. Although **P** is not quantized in whole units of $\hbar$, it is the natural unit for angular momentum.

The Bohr magneton is the natural unit for the magnetic moment of electrons. It can be derived from Eq. (1.5) by substituting $\hbar$ for the magnitude of **P** together with e for the magnitude of the charge on the electron and m_e for its mass. These substitutions give

$$\text{Bohr magneton} = \mu_B = \frac{e\hbar}{2m_e c}. \tag{1.6}$$

A special symbol **L** is used for the orbital angular momentum of an electron in an atom and it is made dimensionless. The defining equation is

$$\mathbf{P}(\text{orbital}) = \hbar\mathbf{L}. \tag{1.7a}$$

It follows from the previous equations that we can write

$$\boldsymbol{\mu}(\text{orbital}) = -\mu_B\mathbf{L}. \tag{1.7b}$$

The minus sign expresses the fact that a rotating electron will have its magnetic moment pointing opposite to the direction of its angular momentum. This is because the electron is negatively charged.

Electrons also possess an intrinsic or "spin" angular momentum, which is independent of their ordinary orbital motion. In classical electricity a charged body that is spun around its own axis will have both an angular momentum and a magnetic moment. The electron must be described in quantum-mechanical terms, and the theory developed by P. A. M.

Dirac (1928) shows that the electron does not necessarily spin about its own axis. The intrinsic angular momentum of the electron comes from its jittermovement (Zitterbewegrug). Instead of traveling in straight lines the electron is forced to have a small-amplitude helical component. It is this motion that gives it an additional angular momentum and magnetic moment. The special symbol **S** is used for this additional angular momentum in dimensionless form, and if we follow the historical practice of calling this a spin angular momentum, we have

$$\mathbf{P}(\text{spin}) = \hbar\mathbf{S}. \tag{1.8a}$$

The γ value for the electron spin is not the same as it is for its ordinary orbital motion, but we can write

$$\boldsymbol{\mu}(\text{spin}) = -g_0\mu_B\mathbf{S} \tag{1.8b}$$

where g_0 is a new physical constant. The interesting point is that g_0 is close to but not exactly equal to 2.

The Dirac theory of the electron predicted that g_0 should be equal to 2, but experimental work showed that it is slightly larger than this value. Although Dirac's work is based upon relativistic quantum mechanics, additional theoretical relativistic corrections have been found. These corrections can be written as a power series in the Sommerfeld fine-structure constant $\alpha = e^2/\hbar c \approx 1/137$. The first term in this power series raises g_0 above 2 by α/π. A commonly quoted value for g_0 is 2.0023. This is accurate to the level of $2 + \alpha/\pi$, but Section A.1 of the Appendix gives the correction to far greater accuracy. For many purposes the Dirac value of 2 is quite satisfactory.

Nuclei can also possess magnetic moments. These magnetic moments are associated with angular momenta, which are again called spin angular momenta. There is even less justification for the use of the word "spin" because nuclear angular momenta are a composite of the intrinsic angular momenta of the nucleons and their orbital angular momenta within the nuclear structure. Both the angular momenta and magnetic moments of nuclei vary a great deal among common nuclei. The nuclear angular momentum **I** is defined in the usual way and

$$\mathbf{P}(\text{nucleus}) = \hbar\mathbf{I}. \tag{1.9a}$$

A quantum number I is associated with the angular momentum **I**. This quantum number is equal to a half or a whole integral number. It is commonly called the nuclear spin or possibly, with a little more accuracy, the nuclear spin quantum number. When $I = 0$ the nucleus has no nuclear magnetic moment, but when $I \geq \frac{1}{2}$ a nuclear moment can be expected. Nuclear moments are measured in terms of a nuclear magneton called μ_N. This is defined similar to the Bohr magneton except that the mass of the proton m_p is used in place of the electron mass m_e in Eq. (1.6). Since the proton is about 1836 times heavier than the electron, μ_N is that much smaller than μ_B.

A nuclear g value can be defined similar to that for electron spin. The defining equation is

$$\boldsymbol{\mu}(\text{nucleus}) = g_I \mu_N \mathbf{I}; \tag{1.9b}$$

however, one finds that g_I is not the commonly tabulated value. Most sources list what is called the "nuclear magnetic moment" and give to it the symbol μ_I. It is simply the product of g_I and the quantum number I, so that the more commonly used expression is

$$\boldsymbol{\mu}(\text{nucleus}) = \frac{\mu_I}{I} \mu_N \mathbf{I}. \tag{1.9c}$$

Section A.3 lists a number of important nuclei together with values for I and μ_I.

Molecular rotation is a source of angular momenta and of rotational magnetic moments which are close to the nuclear magneton. They are particularly important for molecules in the gas phase and for molecules that do not have any net orbital or electron spin moments. They will be

Table 1.1. *Magnetic Moments in Molecules*

Permanent moment

angular momentum $\uparrow \mathbf{P}$ magnetic moment $\uparrow \boldsymbol{\mu} = \gamma \mathbf{P}$

Source	Magnitude	Typical examples
Orbital electron	μ_B	O atom, NO, Ce^{+3}
Electron spin	μ_B	H atom, O_2, Fe^{+3}
Nuclear spin	μ_N	^3He, H_2, HF
Molecular rotation	μ_N	H_2(gas), H_2O(gas)

μ_B = Bohr magneton = 9.2732×10^{-21} erg (gauss)$^{-1}$

μ_N = nuclear magneton = 5.0505×10^{-24} erg (gauss)$^{-1}$

Field-induced moment

field $\uparrow H$ induced moment (often negative) $\uparrow \boldsymbol{\mu} = \alpha \mathbf{H}$

$\alpha(H_2) = -6.6 \times 10^{-30}$ erg (gauss)$^{-2}$

discussed in Chapter 5. The nuclear moments also play their most important role as the moment for nuclear magnetic resonance spectroscopy; this is discussed in Chapter 7.

There is one additional magnetic moment that is important in determining the energy of molecules in a magnetic field. This is the field-induced molecular moment, which is responsible for molecular diamagnetism. Molecules do not possess this moment in the absence of a magnetic field, and this is in contrast with the other moments, which are present even without the field being applied. Table 1.1 is a summary of the various molecular moments.

The field-induced moment is small, and even if 10^5 gauss is present it is still smaller than μ_N. We will find that for bulk samples, in thermal equilibrium, the field-induced moment gives a contribution that is not affected by temperature, whereas the permanent moments are almost averaged away by the thermal equilibrium. The magnetic properties of bulk samples are largely determined by the orbital electron, electron spin, and field-induced molecular moments.

1.2. Bulk Magnetic Properties

Most physical measurements are not made upon individual molecules but are, instead, made upon bulk samples that represent a collection of many molecules. The properties of the bulk samples represent the sums of the properties of the individual molecules; there are, however, at least two extreme types of bulk samples. If the magnetic moments of the individual molecules do not interact strongly, the total magnetic moment of a bulk sample is a result of nearly random molecular moments. This situation is highly desired in theory, for it allows us to neglect interactions between molecules and to determine directly molecular properties from bulk measurements.

If, on the other hand, the magnetic moments of the neighboring molecules interact strongly, the bulk properties will depend strongly upon this interaction, and a study of bulk properties becomes as much a study of molecular interactions as a study of the properties of individual molecules. In solid-state chemistry those solids with strongly interacting magnetic moments can be considered as having either ferromagnetic or antiferromagnetic ordering. The ferromagnetic ones are those in which the individual molecular moments add, and the antiferromagnetic ones are those in which they subtract. The most famous example of ferromagnetism is the loadstone. This naturally occurring ferromagnet was a source of mystery to man for thousands of years, and much of the history of magnetism centers around it.

Since we are primarily interested in the magnetic properties of individual molecules, we shall concentrate our attention away from the ferromagnetic and antiferromagnetic states, although most systems of magnetic moments will be ordered at sufficiently low temperatures. If the individual molecular magnetic moments do not interact, it is clear that such bulk samples will have no permanent magnetic moment. In the absence of an external magnetic field the individual magnetic moments will point all ways in space and their net value will be zero. The contrast among the ferromagnetic, antiferromagnetic, and noninteracting states is shown in Fig. 1.3.

When a collection of noninteracting magnetic moments is placed in a magnetic field, a net magnetic moment will be induced in the bulk sample.

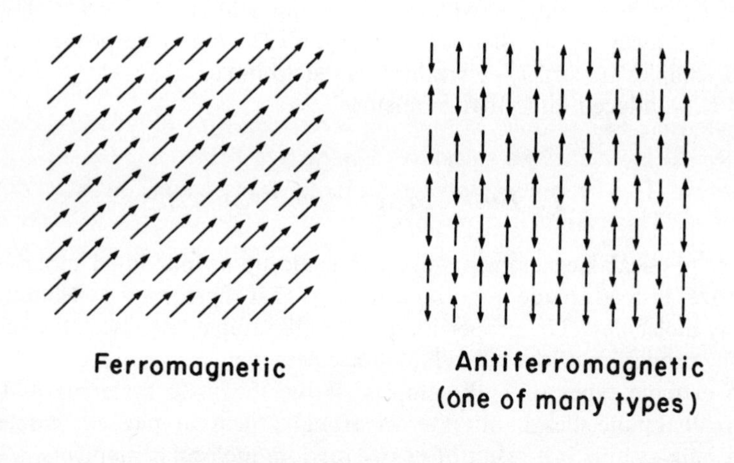

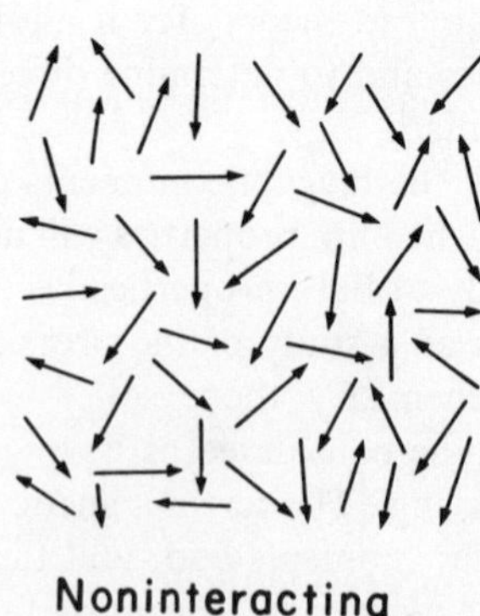

Figure 1.3. *Magnetic states shown for a simple two-dimensional lattice.*

For ordinary magnetic fields this induced moment is directly proportional to the applied field; this is summarized by the equation

$$\mathbf{M} = \chi \mathbf{H}. \tag{1.10}$$

The net induced magnetic moment per unit mass is given the symbol **M**, and χ is called the mass magnetic susceptibility. The susceptibility per mole of material is given the symbol χ_M. The induced moment **M** is a vector, as **H** is also a vector, and it is often called the magnetization vector.† Since in solids **M** and **H** do not have to point in the same direction, χ for solids is a matrix or second-rank tensor which transforms **H** into **M**. In polycrystalline solids one would measure the average susceptibility along the field. This average susceptibility corresponds to the average of the three diagonal elements, or one-third of the trace of the χ matrix.

For samples containing molecules with net electron spin moments, χ is relatively large and always positive in sign. This case is called paramagnetism. For samples containing molecules with only field-induced molecular moments and a negative α value (see Table 1.1), χ is small and negative. This is called diamagnetism. In a few cases the α value is positive. This results in a positive value of χ, and this can be classified as second-order paramagnetism or temperature-independent paramagnetism (T.I.P.).

In the case of true paramagnetism the temperature dependence follows the well-known Curie law, where

$$\chi = \frac{C}{T} \quad \text{or} \quad \frac{1}{\chi} = \frac{T}{C} \tag{1.11}$$

and C is the Curie constant. For solids that are ferromagnetic or antiferromagnetic at low temperature the Curie–Weiss law is more accurate where

$$\chi = \frac{C}{T + \Delta} \quad \text{or} \quad \frac{1}{\chi} = \frac{T}{C} + \frac{\Delta}{C}. \tag{1.12}$$

The temperature Δ is the Curie–Weiss constant. If Δ is negative, it is obvious from the form of Eq. (1.12) that χ becomes very large as T approaches the magnitude of Δ. This is the situation for ferromagnetic samples. The critical temperature for ferromagnetism is called the Curie

† In many books **M** represents the induced moment per unit volume. If this is done, the susceptibility is dimensionless, but in our case χ and χ_M do have dimensions (see Section 1.6).

temperature, and if Eq. (1.12) holds exactly, $-\Delta$ is the Curie temperature. Below the Curie temperature Eq. (1.10) is meaningless, because the sample can have a zero-field moment.

When Δ is positive the sample is antiferromagnetic. The critical temperature for antiferromagnetism is called the Néel temperature. The Néel temperature can be less than the value of Δ by as large as a factor of 5. For our purposes we shall be primarily interested in those substances which have small values of Δ, and we shall be most interested in their magnetic properties above the Curie and Néel temperatures.

Diamagnetic substances have a susceptibility that is nearly independent of temperature, and the molar susceptibility also varies little from the liquid to the solid phase. In Table 1.2 we give a list of representative

Table 1.2. *Some Representative Molar Magnetic Susceptibilities*[a]

Compound	State	$\chi_M \times 10^6$	Compound	State	$\chi_M \times 10^6$
H_2	g	−4.0	CCl_4	l	−66.8
He	g	−1.9	CBr_4	l	−93.7
N_2	g	−12.0	C_6H_6	l	−54.85
NO	g	1,472 (292°)	C_6H_{12}	l	−66.1
	g	2,324 (147°)	$[Cr(H_2O)_6]Cl_3$	s	5,950
O_2	g	3,450	$MnCl_2{\cdot}4H_2O$	s	14,600
H_2O	g	−13.1 (373°)	$FeCl_3{\cdot}6H_2O$	s	15,250
	l	−12.97	$K_3Fe(CN)_6$	s	2,290
	l	−12.93 (273°)	$FeCl_2{\cdot}4H_2O$	s	12,900
	s	−12.65 (273°)	$K_4Fe(CN)_6$	s	−130
KCl	s	−39.0	$CuSO_4{\cdot}5H_2O$	s	1,460
	soln	−39.4	$ZnSO_4{\cdot}7H_2O$	s	−143
KBr	s	−49.1	$Ce(NO_3)_3{\cdot}5H_2O$	s	2,310
KI	s	−63.8	$Ce(SO_4)_2{\cdot}4H_2O$	s	−97
Hg	l	−33.4	$Gd(C_2H_5SO_4)_3{\cdot}9H_2O$	s	24,100
Na	s	+16.1	$U(C_2O_4)_2{\cdot}5H_2O$	s	3,760

[a] Values are for 290–295°K unless otherwise indicated. They are taken from Tables of Constants and Numerical Data (U.I.C.P.A.), No. 7, *Constantes Sélectionnées Diamagnétisme et Paramagnétisme* by G. Foëx, Masson and Cie, Paris (1957). They are in cgs units (see Section 1.6), and the listed numbers correspond to 10^{-6} cm^3 mole^{-1}.

molar susceptibilities for substances either in the gas phase where $\Delta = 0$ or in solids with widely spaced ions so that Δ is close to zero.

It can be seen from Table 1.2 that most molecules possessing an even number of electrons are diamagnetic. The oxygen molecule is an important exception to this rule and the NO molecule possessing an odd number of electrons is also paramagnetic. An interesting fact about NO is that it does not quite follow the Curie law, as can be seen from the figures given. At 292°K its susceptibility has dropped by less than the factor of 2 predicted by Eq. (1.11) over its value at 147°K. The reasons for this will be discussed later.

The various values for H_2O clearly show that diamagnetism can be nearly independent of both temperature and phase. The KCl value in aqueous solution must be characteristic of K^+ and Cl^- ions, and the solid-state value is very similar. The bromide ion and the I^- have increasingly larger susceptibilities, indicating greater α values (see Table 1.1) for the larger ions. Owing to the conduction electrons, metallic sodium has a small paramagnetism that does not follow the Curie law. Because of the Fermi–Dirac statistics of these electrons their paramagnetism falls under the characteristics of second-order paramagnetism and is nearly independent of temperature.

The transition-metal and lanthanide ions form some very striking paramagnetic systems. An interesting case is Fe^{+2}, which has an even number of electrons. In the hydrated ion it is paramagnetic but its cyanide complex is diamagnetic. For Fe^{+3}, with an odd number of electrons, the formation of a cyanide complex decreases the paramagnetism to a level close to that of Cu^{+2} and Ce^{+3}. The lanthanide ion, Gd^{+3}, has a particularly large value for its paramagnetic susceptibility.

The data in Table 1.2 can be summarized by some simple generalizations. Most molecules have an even number of electrons and are probably diamagnetic. A few molecules have an odd number of electrons and are paramagnetic. Diamagnetism is related to molecular structure, but it is little affected by changes in state. A large group of paramagnetic systems are formed by the transition-metal and lanthanide ions. As we shall see later, most of these rules follow from the Pauli principle for the allowed states of electrons.

In Chapters 1 through 4 we will only consider the origins of paramagnetism. The theory of diamagnetism will be considered in Chapter 5.

1.3. Classical Paramagnetism

Many of the features of paramagnetism can be explained by the application of classical Boltzmann statistics. It can be seen from Fig. 1.3 that in a paramagnetic system with zero applied field the moments point equally in all directions. When a field is applied, each moment will have a slightly smaller energy if it points parallel to the field rather than against it. The statistics of these small energy differences yields the Curie-law dependence of paramagnetism.

The orientation energy of each moment $\boldsymbol{\mu}$ in a field $\mathbf{H}$ is

$$\text{energy} = \epsilon = -\boldsymbol{\mu} \cdot \mathbf{H}. \tag{1.13a}$$

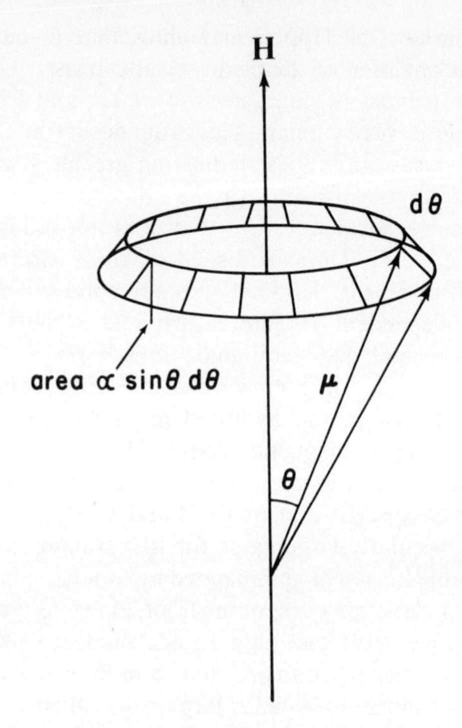

Figure 1.4. *Equal-energy orientations of* $\boldsymbol{\mu}$ *around* **H**. *The number of possible orientations is proportional to* $\sin\theta\, d\theta$.

If the field defines the Z axis, as shown in Fig. 1.4, then we can write

$$\epsilon = -\mu H \cos\theta. \tag{1.13b}$$

In classical mechanics all θ values from 0 to π are possible and the orientation energy can range continuously from $-\mu H$ to $+\mu H$. We shall show that this is not true in quantum mechanics, but if we want to use classical statistics, this full range of energy values is possible. Boltzmann's relation states that these energies are not all equally probable. If we combine Boltzmann's relation with the orientation probability, shown in Fig. 1.4, the fraction of the total number of dipoles at an angle θ is given by

$$\frac{dN(\theta)}{N} = \frac{e^{\mu H \cos\theta/kT} \sin\theta\, d\theta}{\int_0^\pi e^{\mu H \cos\theta/kT} \sin\theta\, d\theta}. \tag{1.14}$$

The Z component of each dipole moment is $\mu \cos \theta$ and the net magnetic moment in the direction of the field is M_Z, so that

$$M_Z = \int \mu_Z \, dN = \frac{N \int_0^{\pi} \mu \cos \theta e^{\mu H \cos \theta / kT} \sin \theta \, d\theta}{\int_0^{\pi} e^{\mu H \cos \theta / kT} \sin \theta \, d\theta} . \tag{1.15a}$$

The integration in Eq. (1.15*a*) is not too difficult, but because $\mu H \ll kT$ above a few degrees Kelvin and at ordinary magnetic fields, the exponentials can be approximated by a power series. The leading term in this power series is the only important one when $\mu H \ll kT$, and Eq. (1.15*a*) gives

$$M_Z = \frac{N\mu^2 H}{3kT} . \tag{1.15b}$$

For 1 mole of dipoles, χ_M follows from Eq. (1.10) and

$$\chi_M = \frac{N\mu^2}{3kT} . \tag{1.16}$$

This result is called the Langevin–Debye equation and it is equally valid for either electric moments in electric fields or magnetic moments in magnetic fields.

The field-induced molecular moment can be very easily included in Eq. (1.16). This moment is equal to αH and its contribution to M_Z is $N\alpha H$. As a result, the field-induced molecular moment simply adds a term equal to $N\alpha$ to χ_M. Since α is usually negative (diamagnetism), the effect of $N\alpha$ is usually to decrease χ_M below the value predicted by Eq. (1.16). One can also see from Eq. (1.15*b*) that the effect of Boltzmann's relation is to decrease the relative importance of the permanent dipoles by a factor of $\mu H/3kT$. It is this factor that makes the field-induced molecular moment relatively more important.

If we neglect the $N\alpha$ contribution to χ_M, Eq. (1.16) can be used to calculate χ_M from μ. For $\mu = \mu_B$ and $T = 293°$K this equation predicts $\chi_M = 427 \times 10^{-6}$ cm^3/mole. A glance at Table 1.2 shows that most paramagnetic substances have larger χ_M values and that μ does not seem to be quantized in whole units of μ_B. The $N\alpha$ contribution cannot account for this discrepancy, for the diamagnetic solids show that $-N\alpha$ should be 150×10^{-6} cm^3/mole or less. A semiempirical constant is often introduced to explain the applicability of Eq. (1.16) to an experimental χ_M value. The effective moment μ_{eff} is defined as the reduced magnetic moment (i.e., $\mu_{\text{eff}} = \mu/\mu_B$), which will give agreement between an observed χ_M and Eq. (1.16). From our calculation (if we neglect $N\alpha$) and with the data in Table 1.2 it can be seen that $(\mu_{\text{eff}})^2$ is $\sim$8 for O_2, $\sim$34 for $MnCl_2 \cdot 4H_2O$, and $\sim$56 for $Gd(C_2H_5SO_4)_3 \cdot 9H_2O$.

These results can be put on a more quantitative basis by a very commonly used formula. If we force Eq. (1.16) to fit an observed χ_M value,

$$\mu_{\text{eff}} = \sqrt{\frac{3kT(\chi_M - N\alpha)}{N\mu_B^2}} = 2.827\sqrt{T(\chi_M - N\alpha)}. \tag{1.17}$$

Since the value of $N\alpha$ is usually small and often unknown, Eq. (1.17) is sometimes written without it.

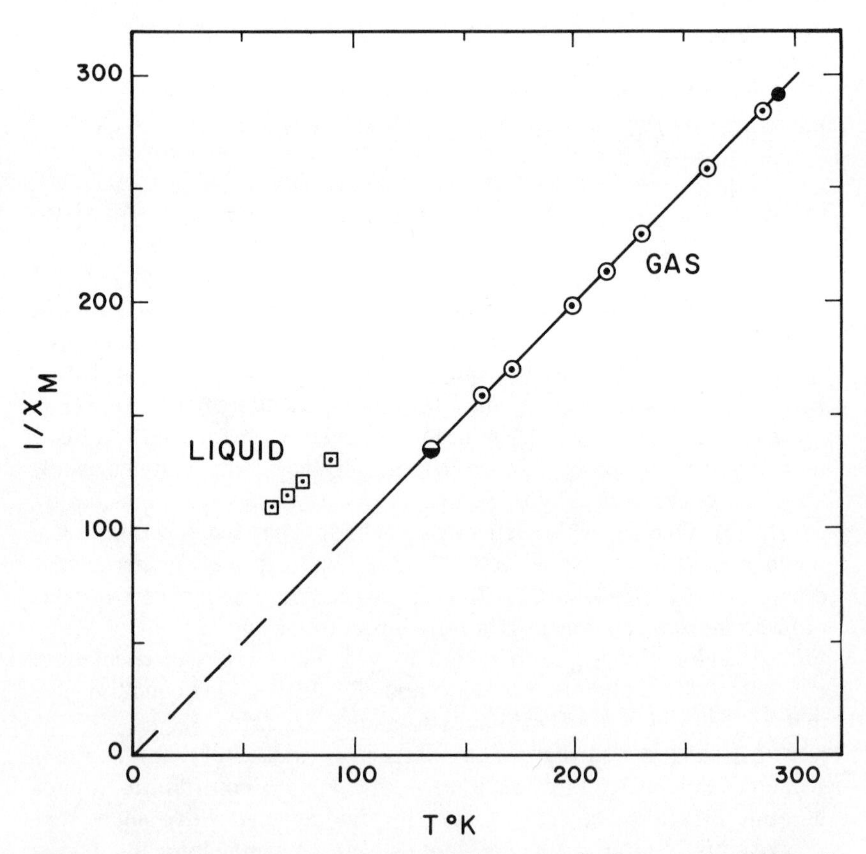

Figure 1.5. *Curie law plot for* O_2. ● *Absolute value E. Bauer and A. Piccard* (1920). ○, ◒ *Relative values of H. R. Woltjer, C. W. Coppoolse, and E. C. Wiersma* (1929) *and R. Sotssel* (1931). (*For details see J. H. Van Vleck.*) *The line is drawn according to Eq.* (1.16) *with* $\mu_{\text{eff}} = \sqrt{8}$. *It is an accident that this predicts a slope of almost exactly* 1.000.

In Fig. 1.5 we show a plot of $1/\chi_M$ vs. T for O_2. The gas nicely follows Eq. (1.16) and it gives $\mu_{\text{eff}} = \sqrt{8}$. The liquid data do not follow the Curie law but approximately follow Eq. (1.12) with $\Delta \sim 50°K$. One must assume that in liquid O_2 the moments interact antiferromagnetically. Some workers have taken this interaction as an indication of the formation of diamagnetic O_4 molecules, but this is probably incorrect. It does show that concentrated paramagnetic liquids may not follow the Curie law because of strong interactions between the moments. In solid O_2 even more striking deviations are obtained for its several crystalline forms.

This value of μ_{eff} for O_2 is a clear indication that angular momentum is not quantized in whole units of $\hbar$. In O_2 the angular momentum is mainly electron spin angular momentum arising from two unpaired electrons, but this is getting ahead of our story. We could at this point introduce the only simple equation that is possible for μ_{eff} in molecules,† but since that equation is only approximate in most systems we shall follow its full development. In NO gas one can see from Table 1.2 that the Curie law is not followed, and one must assume that μ_{eff} varies with temperature. This will be discussed in Chapter 4.

Classical statistics can give only a rough picture of paramagnetism. In Section 1.4 we shall show the actual theoretical ground work based upon quantum statistics. This is an imposing structure with a deep foundation, which is largely the work of J. H. Van Vleck and other workers who pioneered the application of quantum mechanics to magnetic systems. The reader should not be afraid of this theory, for we shall later explain each equation by means of a number of applications.

1.4. Paramagnetic Susceptibilities from Energy Levels

The starting point for any quantum-mechanical calculation is the establishment of the Hamiltonian operator H for the problem. This can be done by first separating H into two parts:

$$\mathsf{H} = \mathsf{H}_0 + \mathsf{H}'(H), \tag{1.18}$$

where H_0 is the Hamiltonian in the absence of the applied magnetic field and $\mathsf{H}'(H)$ is the part with an explicit field dependence. Since the

† For students who want to look ahead it is Eq. (3.7).

energies that arise from the field dependence are small, quantum-mechanical perturbation theory can be used to solve for the result of the addition of $\mathsf{H}'(H)$ to H_0 in Eq. (1.18).

For a paramagnetic substance the major contribution to $\mathsf{H}'(H)$ is the direct interaction of the molecular moment and the field. In classical mechanics this interaction can be simply written as

$$\text{energy} = -\boldsymbol{\mu} \cdot \mathbf{H}. \tag{1.13a}$$

The negative sign in this equation expresses the fact that the lowest possible energy is obtained when $\boldsymbol{\mu}$ and $\mathbf{H}$ are parallel. This orientation places the N pole of a bar magnetic dipole (see Fig. 1.1) closest to the S pole of the magnet that produces the applied field.

The quantum-mechanical Hamiltonian for paramagnetism converts Eq. (1.13*a*) into operator form. Since from Eqs. (1.7*b*) and (1.8*b*) we know the orbital and spin electronic moments in terms of their angular momenta, we can see that

$$\mathsf{H}'(\text{para}) = -\boldsymbol{\mu} \cdot \mathbf{H} = \mu_B H \sum_i [\mathsf{L}_Z(i) + g_0 \mathsf{S}_Z(i)]. \tag{1.19}$$

In this equation we have assumed that $\mathbf{H}$ is entirely along the Z axis and that the electronic orbital and spin moments can be simply summed over all the electrons in the molecule. The diamagnetic contribution to the field-induced molecular moment in Table 1.1 will be accounted for by the addition of a small diamagnetic correction to paramagnetic susceptibilities, and the nuclear spin and molecular rotation moments are too small to be included in Eq. (1.19). The development of H' (para) and H' (dia) is shown in Section A.4 and will be discussed in Chapter 5.

STATISTICS OF PARAMAGNETISM

If Eqs. (1.18) and (1.19) are used in a quantum mechanical calculation, theory allows the resultant energy levels to be expressed as a power series in the field, and for the ith energy level the general form is

$$\epsilon_i = \epsilon_i^0 + \epsilon_i^{(1)} H + \epsilon_i^{(2)} H^2 + \cdots \tag{1.20}$$

where ϵ_i^0, $\epsilon_i^{(1)}$, $\epsilon_i^{(2)}$, etc., are constants independent of the field. The magnetic moment in the direction of the field is given by

$$\mu_Z = -\frac{\partial \epsilon}{\partial H} \tag{1.21}$$

and

$$\mu_{Zi} = -\epsilon_i^{(1)} - 2\epsilon_i^{(2)}H, \tag{1.22}$$

where all higher terms in Eq. (1.20) are neglected.

Equation (1.10) can be used to express χ_M in terms of the number of molecules N_i with the moments μ_{Zi}. For 1 mole,

$$\chi_M = \frac{M_Z}{H} = \frac{1}{H}\sum_i N_i \mu_{Zi} \tag{1.23}$$

and from Eq. (1.22)

$$\chi_M = -\frac{1}{H}\sum_i N_i[\epsilon_i^{(1)} + 2\epsilon_i^{(2)}H]. \tag{1.24}$$

Boltzmann equilibrium determines the number of molecules in each energy level,

$$\frac{N_i}{N} = \frac{e^{-\epsilon_i/kT}}{\sum_i e^{-\epsilon_i/kT}}. \tag{1.25}$$

The terms $\epsilon_i^{(1)}H$ and $\epsilon_i^{(2)}H^2$ are much smaller than kT at most temperatures, so that we can expand the exponentials as

$$e^{-\epsilon_i/kT} = e^{-\epsilon_i^0/kT}(1 - \epsilon_i^{(1)}H/kT \cdots). \tag{1.26}$$

When all these equations are combined in Eq. (1.24), the terms in χ_M that are independent of the applied field are

$$\chi_M = \frac{N\sum_i [\epsilon_i^{(1)}]^2 e^{-\epsilon_i^0/kT}}{kT\sum_i e^{-\epsilon_i^0/kT}} - \frac{2N\sum_i \epsilon_i^{(2)} e^{-\epsilon_i^0/kT}}{\sum_i e^{-\epsilon_i^0/kT}} \tag{1.27}$$

where the zero-field moment is equal to zero and, therefore,

$$\sum_i \epsilon_i^{(1)} e^{-\epsilon_i^0/kT} = 0. \tag{1.28}$$

Our final equation for χ_M looks to be rather complicated and it will require time and study before it can be fully understood. A few generalizations can be made, for energy levels can be expected to follow certain patterns.

One pattern is for some levels to have $\epsilon_i^0 \ll kT$ and for other levels to have $\epsilon_i^0 \gg kT$. In this case the closely spaced levels will give exponentials close to unity, and the widely spaced levels will give exponentials close to zero. If we can set the exponentials equal to 1 or 0, the first term in Eq. (1.27) gives the Curie law and the second term has no temperature dependence. Since we are not including the diamagnetic part of the Hamiltonian in this calculation, the second term must give the temperature-independent paramagnetic part. We will find that $\epsilon_i^{(2)}$ is always negative for the lowest levels and in Eq. (1.22) that $\epsilon_i^{(2)}$ corresponds to an induced moment that is in the same direction as the field.

Another pattern is to have some levels where $\epsilon_i^0 \approx kT$. In this case we would not expect the Curie law to be followed, for the exponentials will then give variations with temperature. Van Vleck refers to "low-frequency" terms as those with $\epsilon_i^0 \ll kT$ and "high-frequency" terms as those with $\epsilon_i^0 \gg kT$. A necessary condition for the Curie law to be followed is that there be no "medium-frequency" terms where $\epsilon_i^0 \approx kT$.

Since the terms $\epsilon_i^{(1)}H$ and $\epsilon_i^{(2)}H^2$ in Eq. (1.20) are usually small compared to ϵ_i^0, they can be calculated by quantum-mechanical perturbation theory. The term $\epsilon_i^{(1)}H$ is the first-order perturbation of the field on ϵ_i^0. It can be calculated from knowledge of the unperturbed wave functions ψ_i^0. The term $\epsilon_i^{(2)}H^2$ is the second-order perturbation of the field on ϵ_i^0, and it results from the mixing of other wave functions ψ_j^0 with ψ_i^0. The amount of this mixing is determined by the difference between the energies $\epsilon_j^0 - \epsilon_i^0$. From perturbation theory we can express $\epsilon_i^{(1)}$ and $\epsilon_i^{(2)}$ by means of the matrix elements of Eq. (1.19) and

$$\epsilon_i^{(1)} = \mu_B \langle \psi_i^0 | \, \mathsf{L}_Z + g_0 \mathsf{S}_Z \, | \psi_i^0 \rangle \tag{1.29}$$

$$\epsilon_i^{(2)} = -\mu_B^2 \sum_{j \neq i} \frac{|\langle \psi_i^0 | \, \mathsf{L}_Z + g_0 \mathsf{S}_Z \, | \psi_j^0 \rangle|^2}{\epsilon_j^0 - \epsilon_i^0} \tag{1.30}$$

where the operators L_Z and S_Z must be summed over all the electrons.

These equations will be used later in the book and we will not try to discuss them further here. However, one important calculation is to relate these quantum-mechanical equations to the classical result given by Eq. (1.16). This can be done fairly easily for one of the energy-level patterns previously discussed. If we assume that all the energy-level spacings are much greater than kT except for the magnetic sublevels which are characteristic of the quantum mechanics of angular momentum, Eq. (1.29) can be used to greatly simplify the first term in Eq. (1.27). It can be seen that the product $-\mu_B(L_Z + g_0 S_Z)$ is equal to the Z component of the magnetic moment, which we call μ_Z. In addition, there is a quantum-mechanical "sum rule" that applies to the magnetic sublevels characteristic of angular momentum. As a result, if we need only sum the first term in Eq. (1.27) over the magnetic sublevels of the ground state, one gets

$$\chi_M = \frac{N \sum_i [\epsilon_i^{(1)}]^2}{kT \sum_i e^{-\epsilon_i^0/kT}} = \frac{N\langle \mu_Z^2 \rangle}{kT}. \tag{1.31a}$$

In this equation $\langle \mu_Z^2 \rangle$ represents the value of μ_Z^2 averaged over the ground state where $\epsilon_i^0 \ll kT$. From vector algebra one knows that $\langle \mu_Z^2 \rangle = \langle \mu_X^2 \rangle = \langle \mu_Y^2 \rangle = \frac{1}{3}\langle \boldsymbol{\mu}^2 \rangle$. If we use this relation, Eq. (1.31a) becomes

$$\chi_M = \frac{N\langle \boldsymbol{\mu}^2 \rangle}{3kT}. \tag{1.31b}$$

This result is obviously very similar to Eq. (1.16). The major difference is that we have now replaced the square of a classical dipole moment μ^2 by $\langle \boldsymbol{\mu}^2 \rangle$, which is the quantum-mechanical average over those states where $\epsilon_i^0 \ll kT$. One should note that we assumed that Eq. (1.30) did not contribute to our final result. We will find that Eq. (1.31b) is more general than we have indicated and that the Curie law can be followed even when both terms in Eq. (1.27) are important. The major assumption for the validity of Eq. (1.31b) is that there are no "medium-frequency" terms where $\epsilon_i^0 \approx kT$.

The "high-frequency" elements in Eq. (1.27) can also be simplified. The summation in Eq. (1.30) is over all the excited states, but when $\epsilon_j^0 - \epsilon_i^0 \gg kT$, the only important energy differences involve an electronic quantum number n and the electronic energies ϵ_n^0. The principle of spectroscopic stability and the "sum rules" for angular momentum allow one to eliminate the sum over the magnetic and angular momentum quantum substates, so

$$\sum_{j \neq i} |\langle \psi_i^0 | \mu_Z | \psi_j^0 \rangle|^2 = \tfrac{1}{3} \sum_{n \neq 0} |\langle \psi_0^0 | \boldsymbol{\mu} | \psi_n^0 \rangle|^2. \tag{1.32}$$

When this is used in Eqs. (1.27) and (1.30),

$$\alpha = \frac{2}{3} \sum_{n \neq 0} \frac{|\langle \psi_0^0 | \boldsymbol{\mu} | \psi_n^0 \rangle|^2}{\epsilon_n^0 - \epsilon_0^0} + \alpha(\text{dia}) \tag{1.33}$$

where the subscript zero has all the "low-frequency" levels included in the ground state.

Through the use of Eqs. (1.27), (1.31*b*), and (1.33) we shall proceed by small steps so that the reader will not have to dwell too long in this section. We hope to bring about an understanding of these general equations by means of a number of applications.

1.5. Experimental Methods

The most common methods for the determination of bulk magnetic susceptibilities depend upon the measurements of a force, or the change in energy with displacement. The change in energy of the individual moments in a bulk sample with field follows from Eq. (1.13*a*) and

$$d\epsilon = -\mu_Z \, dH. \tag{1.34}$$

To obtain the energy E per unit mass of a bulk sample in a field, we must first sum the individual magnetic moments over all the sample. Then we must integrate over the field, because the net moment is an induced moment. These two operations give

$$dE = \sum d\epsilon = -M_Z \, dH = -\chi H \, dH$$

and

$$E = \int_0^H dE = -\tfrac{1}{2}\chi H^2. \tag{1.35}$$

We have assumed both that χ is independent of H and that the internal field is equal to the external field (i.e., χ is small).

FARADAY METHOD

In the Faraday method a small sample is placed in an inhomogeneous field with a large value of dH/dx. The force per unit mass of sample in the x direction is obtained from Eq. (1.35) by

$$F_x = -\frac{dE}{dx} = \chi H \frac{dH}{dx}. \tag{1.36}$$

The pole pieces of the magnet must be specially constructed, and the sample must be placed in an accurately calibrated position in the field.

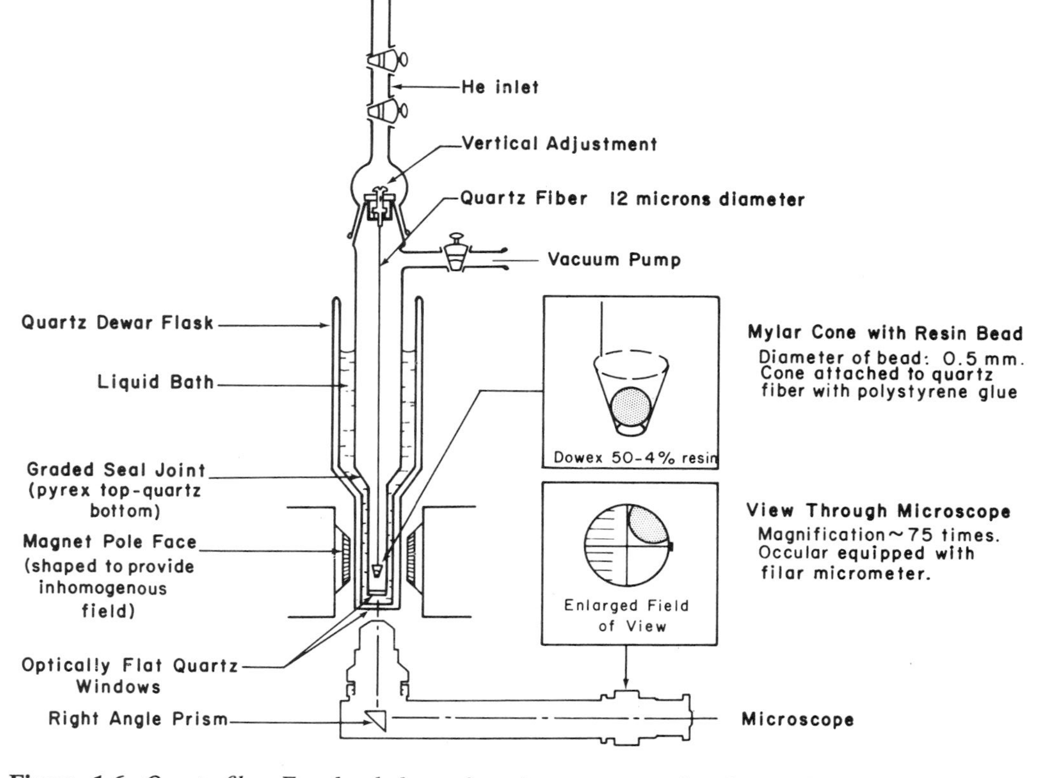

Figure 1.6. *Quartz-fiber Faraday balance for microgram samples.* [*B. B. Cunningham, J. Chem. Ed.*, **36**, 32 (1959).] *The sample is contained in a single bead of an ion-exchange resin, and the gradient in the magnetic field is perpendicular to the plane of the drawing.*

This method is very sensitive and it is also suitable for very small samples. A diagram of an apparatus for microgram samples is shown in Fig. 1.6. Some experimental results from this apparatus are given in Fig. 1.7.

GOUY METHOD

In the Gouy method a long cylindrical sample is suspended so that it extends from the middle to completely outside of the field. This is illustrated in Fig. 1.8. If a small downward displacement were to occur, a small mass of sample, m, is essentially moved from outside the field to the middle of the field. If the tube has an area A and the sample a density

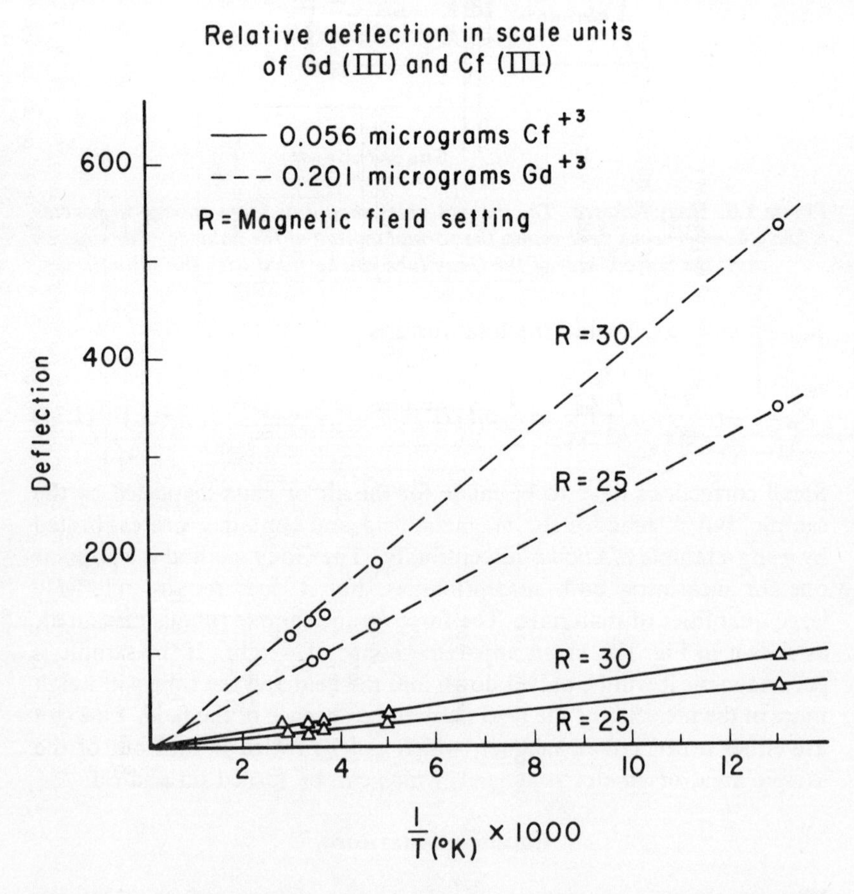

Figure 1.7. *Typical results from the microgram Faraday balance. The* Gd^{+3} *and* Cf^{+3} *were contained on ion-exchange beads. From a knowledge of the susceptibility of* Gd^{+3} *it is possible to calibrate the apparatus.*

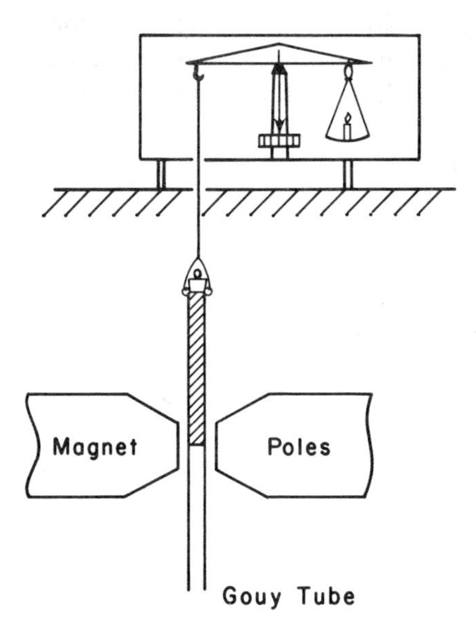

Figure 1.8. *Gouy balance. The magnet poles should be large enough to provide a fairly homogeneous field within the normal motion of the balance. For solution work the bottom half of the Gouy tube can be filled with the solvent.*

ρ, then $m = \rho A \, \Delta x$ and the total force is

$$F_x = -m \frac{\Delta E}{\Delta x} = \frac{m\chi H^2}{2\,\Delta x} = \frac{1}{2}\rho A \chi H^2. \tag{1.37}$$

Small corrections have to be made for the air or glass displaced by the sample, but in practice the magnetic field and container are calibrated by using a sample of known susceptibility. The Gouy method is a popular one for measuring bulk susceptibilities, but it does require relatively large quantities of materials. The force on the sample tube is measured, as shown in Fig. 1.8, as an apparent change of weight. If the sample is paramagnetic it will be pulled down into the field and the tube will weigh more in the presence of the field than in the absence of the field. One can use either a permanent magnet, which can be rolled in and out of the sample area, or an electromagnet, which can be turned on and off.

INDUCTION METHODS

Whereas the two previous methods involve the measurement of the force on a sample in a magnetic field, the inductance of a coil also depends upon susceptibility. If a sample possessing a magnetization **M** is moved along the axis of a coil, a voltage is induced in the coil. This voltage can be used to measure the magnetization of samples on an absolute basis. W. F. Giauque (1962) has shown that very accurate magnetizations can be determined for paramagnetic crystals at temperatures as low as 0.5°K with this induction method.

A popular commercial instrument uses the principle of a vibrating or rotating sample in a magnetic field. It measures the voltage generated in a coil at the frequency of the periodic motion of the sample. It is a sensitive method that is good for small samples over a range of temperatures. Both of these induction methods can take advantage of the large solenoidal fields generated by superconducting magnets.

It is also possible to use alternating magnetic fields and a coil wrapped around a fixed sample. We will find in Chapter 6 that **M**, for a paramagnetic sample, can have a complex frequency dependence, and for this reason only very low-frequency magnetic fields should be used.

MAGNETIC RESONANCE SPECTROSCOPY

When samples containing magnetic moments, arising from either electron spins or nuclear spins, are placed in a magnetic field, a splitting of energy levels is produced. These split levels are spaced rather closely and their differences correspond to frequencies in the radio or microwave portions of the electromagnetic spectrum. After several unsuccessful experiments had been made, electron paramagnetic resonance spectroscopy (EPR) was detected by E. Zavoisky (1945); nuclear magnetic resonance spectroscopy (NMR) was detected also in 1945 by two groups of workers, E. M. Purcell, H. C. Torrey, and R. V. Pound, and F. Bloch, W. W. Hansen, and M. Packard. These techniques will be discussed in some detail in Chapters 7 and 8. Their common features, the fundamentals of magnetic resonance, will be discussed in Chapter 6.

A diagram of a basic magnetic resonance spectrometer is given in Fig. 1.9. It is a completely electronic instrument, but EPR and NMR instruments are very versatile and much research in physical and organic chemistry makes use of these instruments. Section A.3 lists many nuclei that can be detected by NMR. The tabulated Larmor frequencies represent the expected NMR absorption frequencies for an applied field of 10,000 gauss.

Nuclear magnetic resonance frequencies can be measured for liquid samples of only a few tenths of a cubic centimeter to a precision of better than one part per million. Since the fields inside such samples can be changed by a factor close to $(1 + \chi)$, NMR can easily detect such shifts even in ordinary diamagnetic liquids. For this reason, NMR is a fast

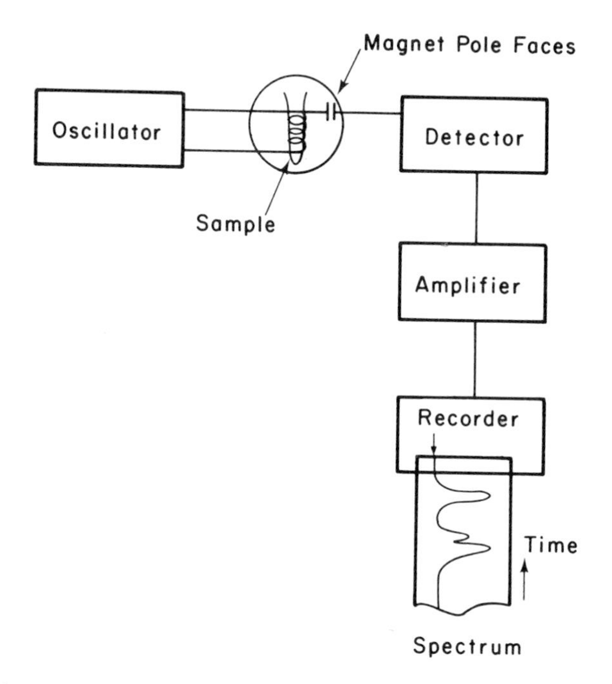

Figure 1.9. *Basic magnetic resonance spectrometer. In NMR the sample is in a coil as shown, but in EPR the higher frequency requires the use of wave guide and the sample is in a resonant cavity. In both techniques the magnetic field is slowly changed with time.*

and convenient way to measure the magnetic susceptibility of liquids or dilute solutions. This technique will also be discussed in Chapter 7.

MOLECULAR BEAMS

If a gas leaks into a vacuum through a series of holes, with dimensions smaller than the mean free path, the gas forms a molecular beam in the vacuum. This beam can be deflected by an inhomogeneous magnetic field if the molecules possess magnetic moments. One of the major experiments in physics was performed by O. Stern and W. Gerlach (1924) when they showed that a beam of silver atoms is deflected into two beams corresponding to two preferred θ values (see Fig. 1.4). This was early solid evidence of the quantization of the projection of angular momentum along the axis of a magnetic field.

It is very difficult, however, to make accurate measurements based upon simple deflection in an inhomogeneous magnetic field, although I. I. Rabi et al. (1934) showed that even nuclear magnetic moments could be detected in deflection experiments. Molecular-beam techniques were carried to their highest accuracy by Rabi in 1937 with the molecular-beam magnetic resonance method. A diagram of this kind of apparatus is shown in Fig. 1.10. In this method magnetic resonance transitions are induced in the *C*-field region. These transitions change the θ value for some molecules in the beam, and these molecules miss the wire detector. Magnetic moments can be measured with this technique to an accuracy of a few parts per million.

The molecular-beam technique is specialized and somewhat limited. The major problems relate to the formulation of a satisfactory detector

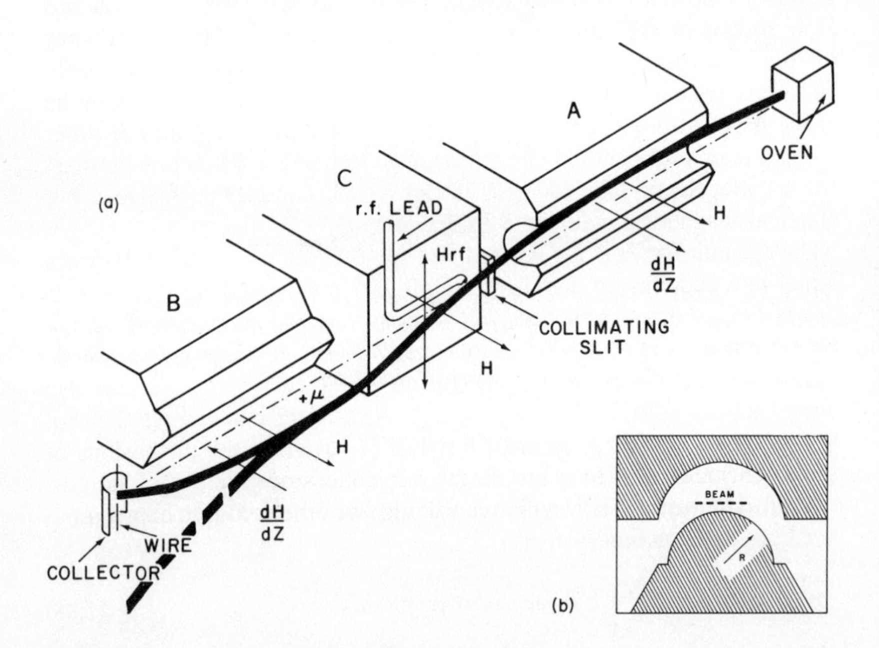

Figure 1.10 *Molecular-beam magnetic resonance apparatus.* [*After P. Kusch, Phys. Today,* **19,** 23 (1966).] *The A and B fields deflect some of the molecules in the beam so that they strike the detector. However, if they change their magnetic states in the C field, because of magnetic transitions induced by the applied radio frequencies, then they follow the dotted path and miss the detector (flop-out technique).* (*a*) *Molecular-beam path in the A, B and C fields.* (*b*) *Cross section of the poles and the beam of the A and B fields.*

for a given molecular beam and to the complicated nature of the apparatus. Nevertheless, many atoms and molecules have been subjected to these investigations, and much fundamental work is possible with this powerful experimental tool.

1.6. Magnetic Units and Dimensions

The equations in this book are written using the single symbol H for either the magnetic induction B or the magnetic field H. This is the common practice of workers in paramagnetism and diamagnetism and we have followed this practice. The use of two magnetic vectors **B** and **H** is rooted in the concept of a homogeneous magnetic material, but on the molecular level it is no longer a necessarily useful distinction. In the cgs system **B** and **H** are equal in a vacuum and are also equal in the space between molecules. In addition, we will find that B and H differ by such a small amount in our materials as to make a distinction between them unnecessary. The use of H in Eqs. (1.10) and (1.35) is correct, but classical magnetism calls for B in Eq. (1.13*a*).

The cgs unit for H is the oersted and for B is the gauss. Since they are equal in a vacuum we use the more attractive word—the gauss. This is again the common practice. The dimensions of both B and H are equivalent to $\text{erg}^{1/2}\ \text{cm}^{-3/2}$ in the cgs system. The magnetic dipole moment μ has dimensions equivalent to $\text{erg}^{1/2}\ \text{cm}^{3/2}$, but one can also write them as gauss cm^3 or erg gauss^{-1}. As a result one can see that Eq. (1.13*a*) gives an energy in ergs, but Eq. (1.16) gives the dimensions of χ_M as $\text{cm}^3\ \text{mole}^{-1}$. These are the correct dimensions for χ_M.

The fundamental relation between **B** and **H** in the common nonrationalized cgs or Gaussian system is

$$\mathbf{B} = \mathbf{H} + 4\pi\mathbf{M}. \tag{1.38}$$

This magnetization **M** is not the same **M** as we used in Eq. (1.10), for the **M** in Eq. (1.38) has the same dimensions as B or H. This is because the magnetization in Eq. (1.38) is the net moment per unit volume and has the dimensions of moment cm^{-3} or $\text{erg}^{1/2}\ \text{cm}^{-3/2}$. The magnetization that we used in Eq. (1.10) has the dimensions of moment g^{-1}, and so our equivalent of Eq. (1.38) would have to include the density. An additional summary of magnetic units is given in Section A.2.

The major advantage of writing all our equations in the old-fashioned cgs system is simplicity. As long as voltage and current do not occur in our equations one does not have the problem of practical versus cgs units. Readers who prefer the more modern mks system can make use of the comparison equations given in Section A.2.

Dimensional analysis of some of the equations in this chapter is given below:

(1.1*b*) $\boldsymbol{\mu}(\text{erg}^{1/2}\ \text{cm}^{3/2}) = i(\text{abampere})A(\text{cm}^2)$

$$1\ \text{abampere} = 10\ \text{ampere (prac.)}$$

(1.2) $$i(\text{abampere}) = \frac{q(\text{esu})\omega(\text{sec}^{-1})}{2\pi c(\text{cm sec}^{-1})}$$

(1.6) $$\mu_B(\text{erg}^{1/2}\ \text{cm}^{3/2}) = \frac{e(\text{esu})\hbar(\text{erg sec})}{2m_e(\text{g})c(\text{cm sec}^{-1})}$$

(1.10) $\mathbf{M}(\text{erg}^{1/2}\ \text{cm}^{3/2}\ \text{g}^{-1}) = \chi(\text{cm}^3\ \text{g}^{-1})\mathbf{H}(\text{erg}^{1/2}\ \text{cm}^{-3/2})$

(1.13*a*) $\epsilon(\text{erg}) = -\boldsymbol{\mu}(\text{erg gauss}^{-1}) \cdot \mathbf{H}(\text{gauss})$

$$= -\boldsymbol{\mu}(\text{erg}^{1/2}\ \text{cm}^{3/2}) \cdot \mathbf{H}(\text{erg}^{1/2}\ \text{cm}^{-3/2})$$

(1.16) $$\chi_M(\text{cm}^3\ \text{mole}^{-1}) = \frac{N(\text{mole}^{-1})[\mu(\text{erg}^{1/2}\ \text{cm}^{3/2})]^2}{3k(\text{erg deg}^{-1})T(\text{deg})}$$

(1.37) $F_x(\text{dyne}) = \frac{1}{2}\rho(\text{g cm}^{-3})A(\text{cm}^2)\chi(\text{cm}^3\ \text{g}^{-1})[H(\text{erg}^{1/2}\ \text{cm}^{-3/2})]^2$

If the reader understands the simplicity of the cgs system in our equations, he should have no difficulty obtaining numerically correct answers.

References

P. Kusch, "The Electron Dipole Moment—A Case History," *Phys. Today*, **19,** 23 (1966). An interesting account of the determination of g_0 for the electron.

R. E. Powell, "Relativistic Quantum Chemistry," *J. Chem. Ed.*, **45,** 558 (1968). A simplified account of the results of Dirac's theory of electrons.

N. F. Ramsey, *Nuclear Moments*, John Wiley and Sons, Inc. New York (1953). The fundamentals of the various methods for the determination of the magnetic moments of nuclei.

P. W. Selwood, *Magnetochemistry*, John Wiley & Sons, Inc. (Interscience Division), New York (1956). The standard modern reference but containing little or no theory.

E. C. Stoner, *Magnetism and Matter*, Methuen and Co. Ltd., London (1934); *Magnetism*, Methuen and Co. Ltd., London (1948). Theory and experiment and some interesting history.

J. H. Van Vleck, *The Theory of Electric and Magnetic Susceptibilities*, Oxford University Press, Inc., New York (1932). The classic work for theory, now reprinted with paper covers.

References for magnetic resonance spectroscopy are given at the ends of Chapters 6, 7, and 8.

Problems†

1.1. Calculate the frequency of rotation $\omega/2\pi$ for an electron with $r = 0.5 \times 10^{-8}$ cm and with $P = \hbar$. Compare this with the Larmor, or precessional, frequency of an electron spin in a field of 10^4 gauss as given in Section A.1. Which is larger?

1.2. Determine the ratio $\mu_B H/kT$ for a field of 10^4 gauss and 300°K. At what temperature will $\mu_B H = kT$ at a field of 10^4 gauss?

1.3. Determine μ_{eff} for NO gas at 292°K and 147°K from the data in Table 1.2. (*Note*: Fig. 4.8 shows the variation of μ_{eff} vs. T for NO in some detail.)

1.4.* Use Eq. (1.27) to determine the χ_M that results from a simple two-energy-level system. Assume two possible levels with $\epsilon = \epsilon^0 \pm \mu_B H$ and determine the χ_M that results. Compare this χ_M with that determined by Eq. (1.16) and evaluate μ_{eff} for this simple two-level system.

1.5. Calculate the force on a 10^{-2}-g sample of $MnCl_2 \cdot 4H_2O$ in a Faraday balance with $H = 10^4$ gauss and $dH/dx = 10^3$ gauss cm^{-1}. If x represents a change in height, convert this force into a change in weight for the sample. Do the same calculation for 10^{-2} g of KCl. What difference other than the size of the weight change would be observed in the two cases?

1.6. Correct Eq. (1.37) for the fact that air is normally displaced from the field when the sample mass m moves into the field. In other words, the complete process is that of a sample mass being moved into the field and an equal volume of air being moved out. If air is 20 mole% O_2 and 80 mole% N_2, determine the percentage of error for a sample of pure water in a Gouy balance if Eq. (1.37) is not corrected for the displaced air at 20°C. (*Note:* The Gouy tube design in Fig. 1.8 automatically corrects for the susceptibility of the glass container but not for the air, unless the lower half is filled with solvent.)

1.7.* Determine the change in weight on a sample of pure water in a Gouy balance. Assume that the tube has an internal diameter of 1.0 cm and that $H = 10^4$ gauss at the center of the field. If the water fills the top half of the Gouy tube in Fig. 1.8, will the water increase or decrease its weight when the field is turned on?

1.8. If electron paramagnetic resonance spectroscopy (EPR) is done on the substance in Problem 1.4, absorption of the oscillator frequency ν in Fig. 1.9 will occur when $h\nu = 2\mu_B H$. If $H = 10^4$ gauss, determine the value of ν. Compare this value with the Larmor frequency of electron spins for 10^4 gauss as given in Section A.1. If you find that these two frequencies are identical,

† The solutions of those problems marked with an asterisk are given at the end of the Problems section in each chapter.

perhaps this energy-level system and the μ_{eff} determined in Problem 1.4 are both characteristics of isolated electron spins.

Solutions:

1.4. We will take $\epsilon_a = \epsilon^0 - \mu_B H$ and $\epsilon_b = \epsilon^0 + \mu_B H$ from Eq. (1.20):

$$\epsilon_a^0 = \epsilon_b^0 = \epsilon^0$$

$$\epsilon_a^{(1)} = -\mu_B \qquad \epsilon_a^{(2)} = 0$$

$$\epsilon_b^{(1)} = \mu_B \qquad \epsilon_b^{(2)} = 0.$$

If we combine these with Eq. (1.27),

$$\chi_M = \frac{N(\mu_B^2 e^{-\epsilon^0/kT} + \mu_B^2 e^{-\epsilon^0/kT})}{kT(e^{-\epsilon^0/kT} + e^{-\epsilon^0/kT})}$$

$$= \frac{N\mu_B^2}{kT}.$$

If we start with Eq. (1.16),

$$\chi_M = \frac{N\mu^2}{3kT},$$

and with our previous result,

$$\frac{N\mu^2}{3kT} = \frac{N\mu_B^2}{kT},$$

or

$$\mu = \mu_B\sqrt{3},$$

or

$$\mu_{\text{eff}} = \sqrt{3}.$$

1.7. For a Gouy balance,

$$\text{force} = \tfrac{1}{2} \times 1.0 \times \tfrac{1}{4}\pi(1.0)^2 \times 13.0 \times 10^{-6} \times 10^8$$
$$= 5.1 \times 10^2 \text{ dyne},$$

where we have used the correct cgs units.

To convert to a change in weight we divide by the acceleration due to gravity,

$$\Delta W = \frac{5.1 \times 10^2}{980} = 0.53 \text{ g wt}$$

Since water is diamagnetic, it will be pushed out of the field. This is equivalent to a decrease in weight.

2

Atomic Structure and Atomic Magnetism

THE ELECTRONIC THEORY of atoms and ions is much easier than the theory of molecules. There is no fundamental difference between the two systems, but in atoms the forces are mainly central forces and, therefore, simple approximate calculations are much more valid for atoms than for molecules. Also, the situation with regard to angular momentum is much simplified in atoms, and since magnetic properties are related to angular momentum, atomic magnetism is a good starting point for a consideration of molecules.

In quantum mechanics, properties such as energy and angular momentum are related to the appropriate operators. These operators are formulated by certain rules that convert the classical relations into quantum-mechanical operators. Once an operator G is established, its eigenfunctions ψ_i and eigenvalues g_i can be determined by the fundamental equation of quantum mechanics,

$$\mathsf{G}\psi_i = g_i\psi_i. \tag{2.1}$$

The Hamiltonian operator H is of special importance for the true wave functions Ψ_i, and the allowed energy levels ϵ_i are determined by the same eigenvalue equation,

$$\mathsf{H}\Psi_i = \epsilon_i \Psi_i. \tag{2.2}$$

If an operator G also has the wave function Ψ_i as its eigenfunctions, then

$$\mathsf{G}\Psi_i = g_i \Psi_i, \tag{2.3}$$

and the dynamical variable represented by G is quantized with the values g_i at the same time that the energy is quantized.

2.1. Angular Momentum and the Atomic Hamiltonian

The Hamiltonian operator for an atom with several electrons can be formulated so that it is the sum of various terms; written in the order of their importance,

$$\begin{aligned}\mathsf{H}(\text{atom}) = {} & \mathsf{H}(\text{kinetic}) + \mathsf{H}(\text{Coulombic}) + \mathsf{H}(\text{spin–orbit}) \\ & + \mathsf{H}(\text{spin–spin}) + \mathsf{H}(\text{nuclear}).\end{aligned} \tag{2.4}$$

The first term simply represents the part that is derived from the classical kinetic energy of the electrons. The second term can be broken into two parts so that

$$\begin{aligned}\mathsf{H}(\text{Coulombic}) &= \mathsf{H}(\text{nuclear–electron attraction}) \\ &\quad + \mathsf{H}(\text{electron–electron repulsion}) \\ &= \mathsf{H}(\text{central force}) + \mathsf{H}(1/r_{ij}).\end{aligned} \tag{2.5}$$

The third term in Eq. (2.4) represents the magnetic interaction of the orbital motion of the electrons with their spin moments, and the fourth term is the direct dipole–dipole interaction between the electron spins. The nuclear term is the least important one. It represents the nuclear magnetic dipole and nuclear electric quadrupole interaction with the electrons. Some nuclei have neither magnetic dipole moments nor electric quadrupole moments, so the last term in Eq. (2.4) would not exist for these nuclei. A very careful consideration of the energy levels of atoms must have additional terms for H(atom), but we shall not use them.

A great deal of information about the properties of electrons in atoms can be gained by using wave functions that are products of one-electron wave functions. If one considers m electrons, an approximate wave function can be written down as the product function

$$\Psi = \psi(1)\psi(2)\psi(3)\cdots\psi(m). \tag{2.6}$$

If one were to select only the first two terms in Eq. (2.4) and only the central force term in Eq. (2.5), the $\psi(i)$ functions would be hydrogen-atom functions with the appropriate nuclear charge Z. These simple hydrogen-atom functions are only a very rough approximation to the real wave functions. An improvement over these functions can be made by introducing some of the effects of electron–electron repulsion into the calculation. This is most easily done by considering the screening of the nuclear charge by the other electrons. The hydrogen-atom wave functions can be altered with different Z(effective) values for each type of electron in an atom. It is then found that p and d electrons are screened more effectively by the other electrons than are s electrons and that they have a higher energy.

With screened orbitals, the electronic structures of most of the elements in the periodic table can be satisfactorily approximated, but the electronic energies predicted by these simple screened functions are not very accurate. D. R. Hartree (1928) developed a self-consistent field technique and it gives the most satisfactory energies. In this technique, electron–electron repulsion is largely taken into account by varying a wave function based upon an assumed charge distribution until it reproduces the charge distribution that was used to calculate the wave function. Because this method is based upon the variation principle of quantum mechanics, the resultant wave function gives the lowest possible energy within the limitations of one-electron wave functions.

All these methods, however, only alter the radial part of the hydrogen-atom wave functions; the angular part remains unchanged. We shall use the symbols $1s$, $2s$, $2p$, etc., to represent the angular part of the hydrogen-atom functions together with the radial part of whichever function gives the best fit to the energy with product wave functions of the type of Eq. (2.6). If this is done, the resultant functions still have the angular momentum properties of the hydrogen-atom functions and

$$\mathbf{L}^2(i)\psi(i) = l(l+1)\psi(i) \tag{2.7a}$$

$$\mathsf{L}_Z(i)\psi(i) = M_l\psi(i) \tag{2.7b}$$

where $l = 0$ for s electrons, $l = 1$ for p electrons, etc., and $M_l = -l, -l+1, \ldots, l$.

To account for the properties of electron spin the latter must be included in the wave function. If the symbols α and β are used for the spin part, we have

$$\mathbf{S}^2(i)\alpha(i) = s(s+1)\alpha(i) = \tfrac{3}{4}\alpha(i) \tag{2.8a}$$

$$\mathbf{S}^2(i)\beta(i) = s(s+1)\beta(i) = \tfrac{3}{4}\beta(i) \tag{2.8b}$$

$$\mathsf{S}_Z(i)\alpha(i) = M_S\alpha(i) = \tfrac{1}{2}\alpha(i) \tag{2.8c}$$

$$\mathsf{S}_Z(i)\beta(i) = M_S\beta(i) = -\tfrac{1}{2}\beta(i). \tag{2.8d}$$

The wave function for each electron $\psi(i)$ will be a product of the spin part and an orbital part, so

$$\begin{aligned}\psi(i) &= \psi(\text{spin}) \times \psi(\text{orbital}) \\ &= \psi(M_S) \times \psi(n, l, M_l)\end{aligned} \tag{2.9}$$

where the quantum numbers n, l, M_l, and M_S specify the state of the electron. In Table 2.1 we illustrate the possible states of the electron suitable for use in Eq. (2.9).

The wave functions in Eq. (2.6) must also satisfy the Pauli principle. This principle requires that no two electrons have all the same quantum numbers. In Table 2.2 we show electron quantum numbers that are

Table 2.1. *States of Electrons in Atoms*

Symbol[a]	l	M_l	M_S	Number of states
ns	0	0	$\pm\frac{1}{2}$	2
np	1	−1	$\pm\frac{1}{2}$	6
	1	0	$\pm\frac{1}{2}$	
	1	1	$\pm\frac{1}{2}$	
nd	2	−2	$\pm\frac{1}{2}$	10
	2	−1	$\pm\frac{1}{2}$	
	2	0	$\pm\frac{1}{2}$	
	2	1	$\pm\frac{1}{2}$	
	2	2	$\pm\frac{1}{2}$	
nf	3	−3	$\pm\frac{1}{2}$	14
	3	−2	$\pm\frac{1}{2}$	
	3	−1	$\pm\frac{1}{2}$	
	3	0	$\pm\frac{1}{2}$	
	3	1	$\pm\frac{1}{2}$	
	3	2	$\pm\frac{1}{2}$	
	3	3	$\pm\frac{1}{2}$	

[a] When $n = 1$, only s; $n = 2$, only s and p; $n = 3$, only s, p, and d; and $n = 4$ only s, p, d, and f.

Table 2.2. *Pauli Principle and Helium-Atom Wave Functions*

$n(1)$	$n(2)$	$l(1)$	$l(2)$	$M_l(1)$	$M_l(2)$	$M_s(1)$	$M_s(2)$	$\sum M_l$	$\sum M_s$
Lowest energy configuration: $\Psi = 1s(i)1s(j)$									
0	0	0	0	0	0	$\frac{1}{2}$	$-\frac{1}{2}$	0	0
0	0	0	0	0	0	$-\frac{1}{2}$	$\frac{1}{2}$		
First excited configuration: $\Psi = 1s(i)2s(j)$									
0	1	0	0	0	0	$\frac{1}{2}$	$\frac{1}{2}$	0	1
1	0	0	0	0	0	$\frac{1}{2}$	$\frac{1}{2}$		
0	1	0	0	0	0	$-\frac{1}{2}$	$-\frac{1}{2}$	0	-1
1	0	0	0	0	0	$-\frac{1}{2}$	$-\frac{1}{2}$		
0	1	0	0	0	0	$\frac{1}{2}$	$-\frac{1}{2}$	0	0
1	0	0	0	0	0	$-\frac{1}{2}$	$\frac{1}{2}$		
0	1	0	0	0	0	$-\frac{1}{2}$	$\frac{1}{2}$	0	0
1	0	0	0	0	0	$\frac{1}{2}$	$-\frac{1}{2}$		
Second excited configuration: $\Psi = 1s(i)2p(j)$									
0	1	0	1	0	± 1	$\frac{1}{2}$	$\frac{1}{2}$	± 1	1
1	0	1	0	± 1	0	$\frac{1}{2}$	$\frac{1}{2}$		
0	1	0	1	0	± 1	$-\frac{1}{2}$	$-\frac{1}{2}$	± 1	-1
1	0	1	0	± 1	0	$-\frac{1}{2}$	$-\frac{1}{2}$		
0	1	0	1	0	± 1	$\frac{1}{2}$	$-\frac{1}{2}$	± 1	0
1	0	1	0	± 1	0	$-\frac{1}{2}$	$\frac{1}{2}$		
0	1	0	1	0	± 1	$-\frac{1}{2}$	$\frac{1}{2}$	± 1	0
1	0	1	0	± 1	0	$\frac{1}{2}$	$-\frac{1}{2}$		
0	1	0	1	0	0	$\frac{1}{2}$	$\frac{1}{2}$	0	1
1	0	1	0	0	0	$\frac{1}{2}$	$\frac{1}{2}$		
0	1	0	1	0	0	$-\frac{1}{2}$	$-\frac{1}{2}$	0	-1
1	0	1	0	0	0	$-\frac{1}{2}$	$-\frac{1}{2}$		
0	1	0	1	0	0	$\frac{1}{2}$	$-\frac{1}{2}$	0	0
1	0	1	0	0	0	$-\frac{1}{2}$	$\frac{1}{2}$		
0	1	0	1	0	0	$-\frac{1}{2}$	$\frac{1}{2}$	0	0
1	0	1	0	0	0	$\frac{1}{2}$	$-\frac{1}{2}$		

allowed by the Pauli principle for the helium atom in its lowest configuration and in its first two excited configurations. The combinations that are shown in pairs in Table 2.2 have all their quantum numbers interchanged between the two electrons. A further application of the Pauli principle shows that these pairs are indistinguishable from one another.

It can be seen from Table 2.2 that only one set of distinguishable quantum numbers is allowed if both electrons are in the $1s$ orbit, but 4 sets are allowed if one electron is in the 1*s* and one in the 2*s*, and 12 sets are allowed if the combinations are the 1*s* and 2*p*. The origins of these sets and the differences among these sets can be explained by angular momenta.

ADDITION OF ANGULAR MOMENTA

One of the results of including the electron–electron repulsion, or $\mathsf{H}(1/r_{ij})$, into the atomic Hamiltonian is that the orbital angular momentum of each individual electron is no longer quantized. It is easy to see, on a logical basis, that once individual electrons repel each other they cannot move independently. Since orbital angular momentum is produced by the ordinary rotation of electrons, Coulombic interaction will lead to a sharing of orbital angular momentum between electrons. This can be put on a more mathematical basis by the use of commutation rules.

If we consider the two operators H and G defined by Eqs. (2.2) and (2.3), it is not difficult to show that

$$\mathsf{HG}\Psi = \mathsf{GH}\Psi,$$

or, in pure operator form,

$$\mathsf{HG} - \mathsf{GH} = 0. \tag{2.10}$$

Since this equation is satisfied for H and G, one says that they commute. Eq. (2.10) can be tested for any two operators, and if it is not satisfied, these operators are said not to commute. If operators do not commute, it is not possible to find a wave function ψ_i that satisfies Eq. (2.1) for both operators.

The significance of $\mathsf{H}(1/r_{ij})$ is that it does not commute with either $\mathsf{L}_Z(i)$ or $\mathbf{L}^2(i)$. As a result, the wave functions of Eq. (2.9) cannot satisfy the Hamiltonian if $\mathsf{H}(1/r_{ij})$ is included. Or, from another point of view, we can say that the true wave functions Ψ_i which satisfy Eq. (2.2) cannot satisfy Eq. (2.7), if $\mathsf{H}(1/r_{ij})$ is included in the Hamiltonian. It is possible to find operators that do commute with $\mathsf{H}(1/r_{ij})$. These operators are related to the total orbital angular momentum of all the electrons.

If we form the operators for the total orbital angular momentum,

$$\mathbf{L} = \mathbf{L}(1) + \mathbf{L}(2) + \cdots + \mathbf{L}(m) \tag{2.11a}$$

$$\mathsf{L}_Z = \mathsf{L}_Z(1) + \mathsf{L}_Z(2) + \cdots + \mathsf{L}_Z(m), \tag{2.11b}$$

it is possible to show that they do commute with $\mathsf{H}(1/r_{ij})$, so

$$\mathsf{H}(1/r_{ij})\mathbf{L}^2 - \mathbf{L}^2\mathsf{H}(1/r_{ij}) = 0$$

$$\mathsf{H}(1/r_{ij})\mathsf{L}_Z - \mathsf{L}_Z\mathsf{H}(1/r_{ij}) = 0.$$

The result of this is that H(atom), including electron–electron repulsion, and L^2 and L_Z can have the same eigenfunctions Ψ with

$$\mathbf{L}^2\Psi = L(L+1)\Psi \tag{2.12a}$$

$$L_Z\Psi = M_L\Psi. \tag{2.12b}$$

We can also add spin angular momenta, so

$$\mathbf{S} = \mathbf{S}(1) + \mathbf{S}(2) + \cdots + \mathbf{S}(m) \tag{2.13a}$$

$$S_Z = S_Z(1) + S_Z(2) + \cdots + S_Z(m), \tag{2.13b}$$

and in a similar way one has

$$\mathbf{S}^2\Psi = S(S+1)\Psi \tag{2.13c}$$

$$S_Z\Psi = M_S\Psi. \tag{2.13d}$$

The quantum numbers L and S form what is called a term, and all the energy levels can be specified by the four quantum numbers L, S, M_L, and M_S plus a running index similar to the principal quantum number.

The result of the addition of $\mathbf{L}(i)$ to form $\mathbf{L}$ and $\mathbf{S}(i)$ to form $\mathbf{S}$ is what is called the Russell–Saunders, or LS, coupling scheme. It is most useful for the lighter elements in the periodic table where H (spin-orbit) is not important. As a result of this scheme the quantum numbers for the individual electrons are not preserved. The total number of states are still given by the number of ways that the one-electron functions can be written, as Table 2.2 illustrates, but the states themselves are specified by the values of L, M_L, S, and M_S. This can be summarized by writing Eq. (2.6) as

$$\Psi = \Psi(L, M_L) \times \Psi(S, M_S).$$

The possible values of L and S can be determined by utilizing the possible values of M_L and M_S. They, in turn, can be determined by Eqs. (2.11b) and (2.13b), so

$$M_L = \sum_i M_l(i)$$

and

$$M_S = \sum_i M_s(i).$$

Once the possible values of M_L and M_S are established, the L and S values can often be most readily deduced from the maximum values for

M_L and M_S. One should always keep in mind, however, the fact that all the states specified by the one-electron quantum numbers must also be specified in the Russell–Saunders coupling scheme. For this reason every one-electron state in Table 2.2 must correspond to a possible state specified by a specific value of L, M_L, S, and M_S.

A term symbolism is used which is similar to the symbol for the hydrogen-atom wave function. When $L = 0$ it is called an S term [Students should try to avoid being confused by the use of a capital S both for the quantum number S in Eq. (2.13) and for the term symbol for $L = 0$.], $L = 1$ is a P term, etc. Since there are $2S + 1$ values for M_S, when $S = 0$ it is called a singlet term, $S = 1$ is a triplet, etc. The value of $2S + 1$ is called the multiplicity. Equations (2.12) and (2.13) are unproved, but in Section 2.2 we shall at least show the basis for their derivation.

We can now look at Table 2.2 and see what kinds of terms are present. For the configuration $(1s)^2$ we have only $M_l(1) + M_l(2) = M_L = 0$ and $M_S(1) + M_S(2) = M_S = 0$ allowed, and this must belong to a term with $L = 0$ and $S = 0$. This is called a singlet S term and is written ^1S.

For the $(1s)(2s)$ configuration we have a maximum M_S value of 1. Since $M_S(\max) = S$ and $M_L(\max) = L$, this configuration must contain an $S = 1$ term or a triplet. Only three of the four states can be accounted for by a triplet, so there must be a second term. It is clear that the two terms for $(1s)(2s)$are ^3S and ^1S.

For the $(1s)(2p)$ configuration $M_L(\max) = 1$ and at the same time $M_S(\max) = 1$ or 0. It is clear that the 12 states must arise from a ^3P and a ^1P term. The ^3P term has $(2L + 1)(2S + 1) = 9$ of the 12 states and the ^1P is the other 3.

The question remains as to how any of the possible states in each configuration are split apart by the elements in H(atom). It is easy to see that if one uses only the central force Coulombic term, all the levels in each configuration will have the same energy. However, if one includes $\mathrm{H}(1/r_{ij})$ into the calculation, a splitting will occur. This splitting is such that it separates the terms from each other, but it produces no splitting within the terms. For the $(1s)(2s)$ configuration of He, it produces a separation of 6421 cm^{-1} between the ^1S and ^3S terms. Since this splitting is between a singlet and triplet state, one is inclined to ascribe it to a magnetic interaction involving the electron spin, but that is not correct.

This splitting between states with different multiplicity, within the same configuration, is due to the Pauli principle in combination with electron–electron repulsion. This principle also states that Ψ must change sign when electrons are interchanged. If this is applied to the wave function including the electron spin part, the following wave functions result from the $(1s)(2s)$ configuration:

$$\frac{\alpha(1)\alpha(2)}{2^{\frac{1}{2}}}[1s(1)2s(2) - 1s(2)2s(1)],$$

$$\frac{\beta(1)\beta(2)}{2^{\frac{1}{2}}}[1s(1)2s(2) - 1s(2)2s(1)],$$

$$\frac{\alpha(1)\beta(2) + \alpha(2)\beta(1)}{2}[1s(1)2s(2) - 1s(2)2s(1)],$$

and

$$\frac{\alpha(1)\beta(2) - \alpha(2)\beta(1)}{2}[1s(1)2s(2) + 1s(2)2s(1)]. \tag{2.14}$$

It can be seen from Eq. (2.14) that the first three functions have the same orbital part. It is also clear from Eqs. (2.8), (2.12), and (2.13) that these are the wave functions of a term with $S = 1$ and that they are the wave functions for the ^3S term. The last function is for the ^1S.

If one examines the two orbital wave functions in Eq. (2.14), one sees that the triplet function has a node when $r(1) = r(2)$, whereas the singlet function does not. This difference between the triplet and singlet function produces a smaller contribution from electron–electron repulsion for the triplet and a lower energy. An old spectroscopic rule of thumb formulated by F. Hund states that the term, within the same configuration, of the highest multiplicity has the lowest energy. The energies for some of the levels of He are given in Table 2.3. It can be observed

Table 2.3. *Some Energy Levels for Helium*[a]

Configuration	Energy (cm^{-1})	Term symbol without J	Term symbol with J
$(1s)^2$	0	^1S	$^1S_0$
$(1s)(2s)$	159,850.318	^3S	$^3S_1$
	166,271.70	^1S	$^1S_0$
$(1s)(2p)$	169,081.111	^3P	$^3P_2$
	169,081.189		$^3P_1$
	169,082.185		$^3P_0$
	171,129.148	^1P	$^1P_1$
$(1s)(3s)$	183,231.08	^3S	$^3S_1$
	184,859.06	^1S	$^1S_0$

[a] Taken from C. E. Moore, *Atomic Energy Levels*, National Bureau of Standards Circular 467, Vol. 1 (1949).

from this table that electron–electron repulsion becomes less important as the electrons occupy orbits with a greater difference in principal quantum number and stay farther apart.

SPIN–ORBIT INTERACTION

The spin–orbit interaction in Eq. (2.4) is of special importance. In terms of energy, it is usually much smaller than is either the kinetic or Coulombic parts of the Hamiltonian, but it is the major interaction term between the spin and orbital angular momenta. For this reason it is of particular importance in the determination of the magnetic properties of both atoms and molecules.

A freely moving electron generates a magnetic field but that field is zero at the position of the electron itself. When the electron is moving at right angles to an electric field, such as one has in an atom, the magnetic field at the electron is not zero. The electron spin moment will obviously interact with such a field. If we consider the interaction of each electron with its own orbital motion, theory shows that

$$\mathsf{H}(\text{spin–orbit}) = \sum_i \xi(r_i)\mathbf{L}(i) \cdot \mathbf{S}(i). \tag{2.15}$$

In an actual calculation this interaction must be averaged over the orbital motion of the electrons. When this is done for LS coupling, the spin–orbit interaction, within a given term, can be reduced to

$$\mathsf{H}(LS) = \lambda \mathbf{L} \cdot \mathbf{S} \tag{2.16}$$

where λ is a constant that can be related to the properties of the individual electrons. In this form one neglects the parts of Eq. (2.15) that give matrix elements between terms, but this is a useful approximation if the terms are widely spaced. The importance of this interaction rests upon the fact that $\mathbf{L} \cdot \mathbf{S}$ and L_Z or S_Z do not commute. This means that the atomic wave function, if one includes spin–orbit interaction, is not an eigenfunction of either L_Z or S_Z. To find an angular momentum that is quantized in the presence of spin–orbit interaction one must add $\mathbf{S}$ and $\mathbf{L}$ to form the total angular momentum,

$$\mathbf{J} = \mathbf{L} + \mathbf{S} \tag{2.17a}$$

$$\mathsf{J}_Z = \mathsf{L}_Z + \mathsf{S}_Z. \tag{2.17b}$$

If one considers $\mathbf{J}$, the important commutation rules are

$$\mathbf{J}^2(\mathbf{L} \cdot \mathbf{S}) - (\mathbf{L} \cdot \mathbf{S})\mathbf{J}^2 = 0$$

$$\mathsf{J}_Z(\mathbf{L} \cdot \mathbf{S}) - (\mathbf{L} \cdot \mathbf{S})\mathsf{J}_Z = 0.$$

As a result, the wave functions including spin–orbit interaction can quantize $\mathbf{J}^2$ and J_Z with

$$\mathbf{J}^2\Psi = J(J+1)\Psi \tag{2.18a}$$

$$J_Z\Psi = M_J\Psi = (M_L + M_S)\Psi. \tag{2.18b}$$

Whereas Eqs. (2.12*b*) and (2.13*d*) are no longer valid for the true wave functions, $\mathbf{L}^2$, $\mathbf{S}^2$, $\mathbf{J}^2$, and J_Z do commute with $\mathbf{L}\cdot\mathbf{S}$ and they are quantized by the set of wave functions that are useful as long as λ in Eq. (2.16) is not too large.

With *LS* coupling the contribution of spin–orbit interaction to the energy can be easily calculated, and from Eq. (2.17*a*)

$$\mathbf{J}^2 = (\mathbf{L}+\mathbf{S})^2 = \mathbf{L}^2 + \mathbf{S}^2 + 2\mathbf{L}\cdot\mathbf{S}$$

and

$$\mathbf{L}\cdot\mathbf{S} = \tfrac{1}{2}(\mathbf{J}^2 - \mathbf{L}^2 - \mathbf{S}^2). \tag{2.19}$$

Combining Eqs. (2.12*a*), (2.13*c*), (2.16), (2.18), and (2.19) we find

$$\epsilon(\text{spin–orbit}) = \frac{\lambda}{2}\,[J(J+1) - L(L+1) - S(S+1)] \tag{2.20a}$$

and the famous Landé interval rule

$$\epsilon(J, L, S) - \epsilon(J-1, L, S) = \lambda J. \tag{2.20b}$$

The result of Eq. (2.20) is that the Russell–Saunders terms are further split according to the total angular momentum. The value of J is determined by the ways that $\mathbf{L}$ and $\mathbf{S}$ can add in Eq. (2.17*a*). One finds that

$$J = L+S, L+S-1, \ldots, |L-S|. \tag{2.21}$$

The allowed J values can also be determined by considering Eqs. (2.17*b*) and (2.18*b*). For the ^3P term in He, for example, the M_L and M_S values in Table 2.2 can be used to evaluate the J values. If J is added as a right subscript one finds that the ^3P term, with spin–orbit interaction, is broken down into $^3P_2$, $^3P_1$, and $^3P_0$ states. The number of states in each of these terms is given by the number of M_J values so that each of these terms is $2J+1$ degenerate. The 9 states in the ^3P term break up, by means of spin–orbit interaction, into terms containing 5, 3, and 1 state each.

The value of λ depends upon the atom and its configuration. For the first few elements in the periodic table, λ is 1 cm^{-1} or less, but for many elements λ is several hundred cm^{-1}, and for the heaviest elements λ can be close to 10,000 cm^{-1}. The small splittings between the ^3P states in He show that λ is very small in this case. Equation (2.20*a*) predicts that the spacing of the ^3P terms should be equal to λ, $-\lambda$, and -2λ for the $J = 2$, 1, and 0 states, respectively. It can be observed from Table 2.3 that the splittings of the ^3P terms in He do not agree with these predictions and that λ would have to be negative. In this case λ is so small that other effects such as relativity serve to split the ^3P terms by a small amount and that spin–orbit interaction is just not very important for He.

The atomic wave functions can be specified as $\Psi = \Psi(L, S, J, M_J)$. Although these wave functions are never the true wave functions of an atom, as long as λ is not zero, they are very useful functions. When spin–orbit interaction becomes as important as the electron–electron repulsion that produces the separation into terms, the *LS* coupling scheme is a poor approximation. In this case, one must average Eq. (2.15) over each electron orbital with the result

$$\mathsf{H}(nl) = \sum_i \zeta(n_i l_i)\mathbf{S}(i) \cdot \mathbf{L}(i). \tag{2.22}$$

In this form the spin–orbit interaction only commutes with the total angular momentum of each electron, $\mathbf{J}^2(i)$ and $\mathsf{J}_Z(i)$. The formation of the atomic wave function as a product of one-electron functions, specified by a principal quantum number together with individual total angular momentum quantum numbers $j(i)$ and $M_j(i)$, is called *jj* coupling. This coupling scheme is most useful for the heavy elements, where spin-orbit interaction has its greatest importance.

When one considers atoms with a number of electrons, the closed electronic shells that occur at the rare gas configurations introduce considerable simplification. For Mg, for example, the first 10 electrons can be considered as forming an $^1S_0$ core with a $(1s)^2(2s)^2(2p)^6$ configuration. Most of the electronic states of Mg can be considered as configurations of the last two electrons, and for comparison with He some energy levels for Mg are given in Table 2.4.

It can be seen that Mg is similar to He except that a *p* orbital of the same principal quantum is available and a low *d* orbital is also available. The heavier atom also has much more closely spaced energies. The Landé interval rule is nicely followed by the lowest ^3P but not by the other ^3P and the ^3D. This is presumably because λ is too small for the later cases, and other effects become important.

The atomic systems with the most interesting magnetic properties are those with open-shell ground-state configurations. Examples of such atoms are carbon $[(1s)^2(2s)^2(2p)^2]$, oxygen $[(1s)^2(2s)^2(2p)^4]$, and $V^{+3}[(Ar)(3d)^2]$. Each of these configurations can yield several terms, but Hund's rule requires the lowest energy term for each to be the one with the highest multiplicity. As a result both carbon and oxygen have ^3P

Table 2.4. *Some Energy Levels for Magnesium*

Configuration	Energy (cm^{-1})	Term symbol
$(3s)^2$	0	$^1S_0$
$(3s)(3p)$	21,850.368	$^3P_0$
	21,870.426	$^3P_1$
	21,911.140	$^3P_2$
$(3s)(3p)$	35,051.36	$^1P_1$
$(3s)(4s)$	41,197.37	$^3S_1$
	43,503.0	$^1S_0$
$(3s)(3d)$	46,403.14	$^1D_2$
$(3s)(4p)$	47,847.7	$^3P_0$, $^3P_1$
	47,851.8	$^3P_2$
$(3s)(3d)$	47,957.035	$^3D_3$
	47,957.018	$^3D_2$
	47,957.047	$^3D_1$
$(3s)(4p)$	49,346.6	$^1P_1$

ground-state terms and V^{+3} has ^3F. For such partially filled orbitals there is a simple relationship between λ in Eq. (2.16) and $\zeta(nl)$ in Eq. (2.22). For the term of highest multiplicity $\lambda = \pm\zeta(nl)/2S$, where the plus sign is for less than half-filled orbitals and the negative sign is for greater than half-filled orbitals. In the case of half-filled orbitals the ground state is an S term and $\lambda = 0$. Since the one-electron parameters $\zeta(nl)$ are always positive, this relation gives positive λ values for the lowest energy terms with less than half-filled orbitals and negative λ values with more than half-filled orbitals. The sign of λ is obviously important if we want to know the J value for the ground state.

2.2. Vector Model for Angular Momentum

A consideration of the quantum mechanics of angular momentum is best done by using the commutation rules for the components of a general angular momentum operator P, which are

$$\begin{aligned} \mathsf{P}_X\mathsf{P}_Y - \mathsf{P}_Y\mathsf{P}_X &= i\hbar\mathsf{P}_Z \\ \mathsf{P}_Y\mathsf{P}_Z - \mathsf{P}_Z\mathsf{P}_Y &= i\hbar\mathsf{P}_X \\ \mathsf{P}_Z\mathsf{P}_X - \mathsf{P}_X\mathsf{P}_Z &= i\hbar\mathsf{P}_Y. \end{aligned} \tag{2.23}$$

This commutation rule is much more specific than we previously considered but, in a general sense, we can say that P_X, P_Y, and P_Z do not commute. If, however, one forms the magnitude of $\mathbf{P}$ by

$$\mathbf{P}^2 = \mathsf{P}_X^2 + \mathsf{P}_Y^2 + \mathsf{P}_Z^2, \tag{2.24}$$

one can show, after a little algebra, that $\mathbf{P}^2$ does commute with P_X, P_Y, and P_Z. This means that Ψ can be eigenfunctions of $\mathbf{P}^2$ and one component of $\mathbf{P}$, but that it can never be an eigenfunction for two components of $\mathbf{P}$.

A detailed analysis of Eq. (2.23) leads to a general set of equations which can be applied to all types of angular momenta. If one chooses to satisfy Eq. (2.23) with the square of the magnitude of an angular momentum and its Z component as quantized, the final results are given by Eqs. (2.7), (2.8), (2.12), (2.13*c*), (2.13*d*), and (2.18). It is for this reason that the eigenvalue equations for orbital angular momentum, spin angular momentum, and the total angular momentum are all so similar. One finds that this same set of equations also can be used for the internal angular momenta of nuclei and for the rotational angular momenta of molecules.

In Chapter 1 we showed that each source of angular momentum contributes to the total magnetic moment according to the equation $\mathbf{M}_i = \gamma_i \mathbf{P}_i$. In classical mechanics $\mathbf{P}$ is a vector that can have fixed components P_X, P_Y, and P_Z. The quantum-mechanical commutation rules, Eq. (2.23), allow only one of these components to be fixed. The other two components must vary with time, so that P_X and P_Y cannot be quantized if P_Z is quantized.

This general result for angular momentum is summarized conveniently in what is known as the vector model. The vector-model diagram for a single angular momentum vector $\mathbf{J}$ is shown in Fig. 2.1*a*. In this picture $\mathbf{J}$ appears to precess around the Z axis. If it did not precess, J_X and J_Y could be quantized and this would violate Eq. (2.23). The vector $\mathbf{J}$ has a length given by $\sqrt{J(J+1)}$ where J is an integer or half-integer and the projection of $\mathbf{J}$ along the Z axis is M_J. An interesting feature of the vector model is that $\mathbf{J}$ can never be perfectly aligned along the Z axis, for M_J is never as large as $\sqrt{J(J+1)}$. This classical model satisfies all the conditions imposed by Eq. (2.23).

In the case of two sources of angular momenta, let us say $\mathbf{L}$ and $\mathbf{S}$, there are two vector-model pictures that satisfy Eq. (2.23). In Fig. 2.1*b* we show $\mathbf{L}$ and $\mathbf{S}$ precessing independently with projections along Z of M_L and M_S. On the other hand, $\mathbf{L}$ and $\mathbf{S}$ could precess together and add to form a total angular momentum $\mathbf{J}$. The only projection along Z in this case would be M_J. This is illustrated in Fig. 2.1*c*. The key to the difference between Figs. 2.1*b* and 2.1*c* is the frequency of precession of the angular momenta. In Fig. 2.1*c* the momenta must precess exactly together. Which model one picks when two angular momenta are present

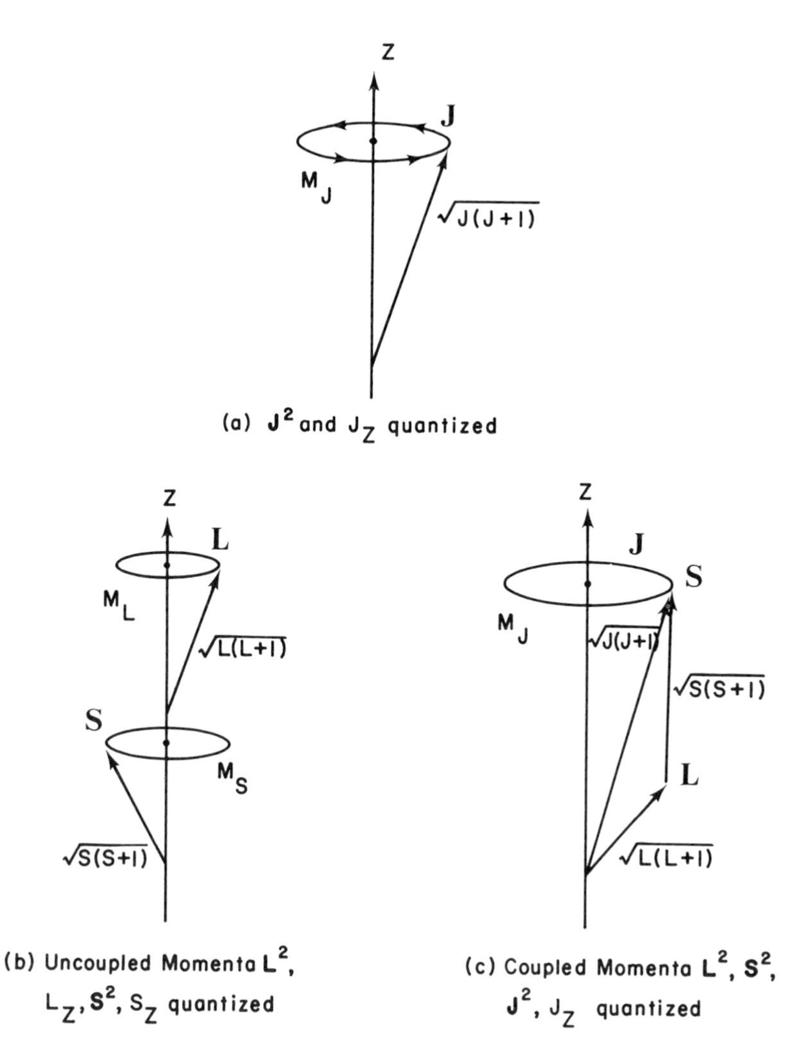

Figure 2.1. *Vector model for angular momentum.* (*a*) $\mathbf{J}^2$ *and* J_Z *quantized.* (*b*) *Uncoupled momenta* $\mathbf{L}^2$, L_Z, $\mathbf{S}^2$, *and* S_Z *quantized.* (*c*) *Coupled momenta* $\mathbf{L}^2$, $\mathbf{S}^2$, $\mathbf{J}^2$, *and* J_Z *quantized.*

depends upon the balancing of the characteristics that tend to make **L** and **S** precess together and those that make them precess separately.

It should be emphasized that Fig. 2.1 and the vector model are not a substitute for good quantum mechanics. They are only a device to help understand the results of quantum mechanics. Particularly, when one is deciding between Fig. 2.1*b* and 2.1*c*, one should always realize that only a real quantum-mechanical calculation will provide the answer and that the final result will almost always lie between the extremes of Fig. 2.1*b* and 2.1*c*.

In order to consider further the two extremes for the two angular momenta cases, let us take an example. In Eq. (2.16) spin-orbit interaction was written as $\mathbf{L} \cdot \mathbf{S}$. If we consider the average value of this interaction, then for Fig. 2.1*b*,

$$(\mathbf{L} \cdot \mathbf{S})_{\mathrm{av}} = M_L M_S, \tag{2.25}$$

and for Fig. 2.1*c*,

$$(\mathbf{L} \cdot \mathbf{S})_{\mathrm{av}} = \tfrac{1}{2}[J(J+1) - L(L+1) - S(S+1)]. \tag{2.26}$$

The last case was used in Eq. (2.19) to obtain Eq. (2.20). We will find several examples of this important difference and the first will be in the Zeeman effect for atoms.

2.3. Zeeman Effect for Atoms

When an atom is placed in a magnetic field, the effect of the field can most easily be detected through changes in atomic spectra. In spectroscopy, this is called the Zeeman effect, in honor of its discoverer, P. Zeeman (1897). It is a very valuable tool in atomic spectroscopy as an aid for assigning spectra; we, however, are interested in the effect of magnetic fields upon atomic energy levels.

When a molecule or atom is placed in a magnetic field we can follow Eq. (1.18), and its Hamiltonian can be written as

$$\mathsf{H} = \mathsf{H}_0 - \boldsymbol{\mu} \cdot \mathbf{H} \tag{2.27}$$

where H_0 is the field-independent part. For a gas-phase atom $\mathsf{H}_0 = \mathsf{H}(\text{atom})$, as given in Eq. (2.4). If the atom possesses a magnetic moment, $\boldsymbol{\mu}$ should be largely field independent. If $\mathbf{J}$ is the total angular momentum of the atom, Eqs. (1.6*b*) and (1.7*b*) can be generalized in operator form with

$$\boldsymbol{\mu} = -g_J \mu_B \mathsf{J} \tag{2.28}$$

If $\mathbf{H}$ defines the Z axis,

$$\mathsf{H} = \mathsf{H}_0 + g_J \mu_B H \mathsf{J}_Z \tag{2.29}$$

and, from Eqs. (2.2) and (2.18*b*),

$$\epsilon = \epsilon_0 + g_J \mu_B H M_J. \tag{2.30}$$

The result of Eq. (2.30) is that the $2J + 1$ degeneracy in each term is now broken, but we do have the problem of the evaluation of g_J.

The magnetic fields that are used are usually in the range 1000 to 100,000 gauss. At these fields $\mu_B H$ represents an energy of 0.05 to 5 cm^{-1}. For many cases g_J is nearly constant and the Zeeman effect is first order (linear) in the field. In some cases the term values are rather closely spaced and g_J is a function of the applied field. In these situations the calculation is more difficult and the Zeeman effect would include terms that are second order (quadratic) in the applied field.

The theory for the Zeeman effect in the case of atoms with only orbital or only spin angular momentum is quite simple. For ^1P, ^1D, etc. states $\mathbf{J} = \mathbf{L}$, so that a comparison of Eqs. (2.28) and (1.6*b*) provides the answer that $g_J = 1$. For ^2S, ^3S, etc. states $\mathbf{J} = \mathbf{S}$, so that a comparison of Eqs. (2.28) and (1.7*b*) provides the answer that $g_J = 2.00232 \approx 2$.

In the case of terms with both spin and orbital magnetic moments, then, the situation is more complicated. In this case, if we combine Eqs. (1.6*b*) and (1.7*b*) in operator form,

$$\boldsymbol{\mu} = -\mu_B[\mathbf{L} + 2.00232\mathbf{S}],$$

but for most purposes we will use

$$\boldsymbol{\mu} = -\mu_B[\mathbf{L} + 2\mathbf{S}]. \tag{2.31}$$

If we compare Eqs. (2.31) and (2.28),

$$g_J\mathbf{J} = \mathbf{L} + 2\mathbf{S},$$

and multiply by $\mathbf{J}$, we obtain

$$g_J\mathbf{J}^2 = \mathbf{L}\cdot\mathbf{J} + 2\mathbf{S}\cdot\mathbf{J}. \tag{2.32}$$

If Eqs. (2.17) and (2.18) are valid, we may use the vector-model diagram in Fig. 2.1*c* or the following algebra:

$$\mathbf{S}^2 = (\mathbf{J} - \mathbf{L})^2 \quad \text{and} \quad \mathbf{L}\cdot\mathbf{J} = \tfrac{1}{2}[\mathbf{J}^2 + \mathbf{L}^2 - \mathbf{S}^2]$$

$$\mathbf{L}^2 = (\mathbf{J} - \mathbf{S})^2 \quad \text{and} \quad \mathbf{S}\cdot\mathbf{J} = \tfrac{1}{2}[\mathbf{J}^2 + \mathbf{S}^2 - \mathbf{L}^2].$$

These results simplify Eq. (2.32), so that

$$\begin{aligned} g_J\mathbf{J}^2 &= \tfrac{3}{2}\mathbf{J}^2 + \tfrac{1}{2}\mathbf{S}^2 - \tfrac{1}{2}\mathbf{L}^2 \\ &= \mathbf{J}^2 + \tfrac{1}{2}(\mathbf{J}^2 + \mathbf{S}^2 - \mathbf{L}^2), \end{aligned} \tag{2.33}$$

and if we use Eqs. (2.11), (2.13), and (2.18),

$$\begin{aligned} g_J &= 1 + \frac{J(J+1) + S(S+1) - L(L+1)}{2J(J+1)} \\ &= \frac{3}{2} + \frac{S(S+1) - L(L+1)}{2J(J+1)} . \end{aligned} \tag{2.34}$$

Equation (2.34) is the well-known Landé – g-value equation for the anomalous Zeeman effect in atomic spectra. This was called the anomalous Zeeman effect because it did not follow the simple relationship with a $g = 1$. The discovery of this effect was strong evidence for the presence of electron spin and its own anomalous g value. Since J, L, and S are constants for each term in LS coupling, Eq. (2.34) predicts g_J values that are constants for every term. They run from about 4 to -1 for the most common terms. If LS coupling is a poor approximation to the real wave functions of the atom, there is a g sum rule that is useful. It can be shown theoretically that the sum of all the g_J values in a given configuration and for a given J value is independent of the actual coupling. Since the sum for LS coupling can be readily calculated, it can be used to check the experimental sum if spin–orbit intraction is large.

When the spin–orbit interaction is small, $\mu_B H$ can become comparable to the spacing between the term values of different J value. In this case the addition of $\mathbf{L}$ and $\mathbf{S}$ becomes a poor approximation and Eq. (2.34) is inaccurate. This results from the fact that $\mathbf{J}^2$ is no longer quantized in the presence of a field. The reasons for this become apparent when in Eq. (2.31) one realizes that $\mathbf{J}^2$ does not commute with $\mathbf{L} + 2\mathbf{S}$, although it does commute with $\mathbf{L}^2$, $\mathbf{S}^2$, and $\mathbf{L} + \mathbf{S}$. In a simple sense the total angular momentum of a molecule is not often quantized in an external field, because the field can act to exchange angular momentum between the laboratory and the molecule.

The effect of the field for small λ is to uncouple **L** and **S** in the vector model so that Fig. 2.1*b* becomes valid. We shall later show that this uncoupling can be simply related to the classical precession frequencies of magnetic moments. Since the g value for the electron spin moment is 2, **S** acts as if it would prefer to precess twice as fast as **L**. The interaction that tends to make **L** and **S** precess together is, of course, λ, the spin–orbit interaction. This uncoupling of **L** and **S** is known as the Paschen–Back effect. In the event that L_Z and S_Z were quantized as shown in Fig. 2.1*b*, then from Eqs. (2.25), (2.27), and (2.31),

$$\epsilon(\text{Paschen–Back}) = \epsilon_0' + \lambda M_L M_S + \mu_B H(M_L + 2M_S) \tag{2.35}$$

where ϵ_0' would be the zero-field energy without spin–orbit interaction.

The transition between Eqs. (2.28) and (2.34) to Eq. (2.35) will involve shifts in the levels that are nonlinear in the field. At low fields the onset of the Paschen–Back effect would be the dependence of $\boldsymbol{\mu}$ upon H and the observation of a second-order term in the Zeeman splittings. If an atom contains a nuclear magnetic dipole moment, its Zeeman effect is quite complicated even at very low fields. This is because the nuclear spin **I** must be added to **J** to form a new total angular momentum, and a Paschen–Back effect to uncouple **I** and **J** can occur at fields of only a few hundred gauss. To illustrate this point, we will consider the Zeeman effect of the hydrogen atom.

2.4. Zeeman Energies for the Hydrogen Atom and the Paschen–Back Effect

The ground electronic state of the hydrogen atom is accurately given by the (1*s*) wave function, and its electronic wave function can be written as

$$\Psi(\text{electronic}) = \alpha(1s) \text{ or } \beta(1s). \tag{2.36}$$

The normal isotope of hydrogen, ^1H, has a nuclear spin of $\frac{1}{2}$, and the total wave function for hydrogen must be a product of nuclear and electronic parts, so that

$$\Psi(\text{total}) = \Psi(\text{nuclear}) \times \Psi(\text{electronic}). \tag{2.37}$$

The nuclear wave function is an eigenfunction of the nuclear spin operators and we can have the usual results for angular momentum,

$$\mathbf{I}^2\Psi(\text{nuclear}) = I(I+1)\Psi(\text{nuclear}) \tag{2.38a}$$

$$\mathbf{I}_Z\Psi(\text{nuclear}) = M_I\Psi(\text{nuclear}) \tag{2.38b}$$

where, for the proton, $I = \frac{1}{2}$ and $M_I = \pm\frac{1}{2}$. The total wave function then has states that can be written as

$$\Psi(\text{total}) = \Psi(M_I) \times \Psi(M_S) \times (1s). \tag{2.39}$$

The total angular momentum is the sum of the nuclear and electronic angular momenta, so that we can define

$$\mathbf{F} = \mathbf{I} + \mathbf{J}, \tag{2.40}$$

and for the (1*s*) state it is a $^2S_{1/2}$ term and $\mathbf{S} = \mathbf{J}$. If Eq. (2.40) is to be used and the states of the hydrogen atom are to be characterized by the total angular momentum, Eq. (2.39) has to be modified as

$$\Psi(\text{total}) = \Psi(F, M_F) \times (1s). \tag{2.41}$$

In the vector model, Eq. (2.39) corresponds to Fig. 2.1*b*, and Eq. (2.41) corresponds to Fig. 2.1*c*. Which wave function one uses depends upon the Hamiltonian.

For the $(1s)$ electronic state the nuclear part of Eq. (2.4) has a very simple form,

$$\mathsf{H}(\text{nuclear}) = A\mathbf{I} \cdot \mathbf{S}. \tag{2.42}$$

This form results from the Fermi contact interaction of the $(1s)$ wave function and the magnetic moment of the proton. The direct dipole–dipole interaction between the electron moment and the nuclear moment is zero because of the spherical symmetry of the $(1s)$ function. The constant A is rather large for a nuclear interaction and is equal to 0.04738 cm^{-1}. The rest of the terms in Eq. (2.4) do not depend upon either the spin of the electron, for the $(1s)$ state of hydrogen, or the spin of the proton. Equation (2.42) is the only part that depends upon $\mathbf{I}$ or $\mathbf{S}$, and so we can call it the "spin Hamiltonian" for zero field, where the rest of Eq. (2.4) is just an uninteresting constant—from the point of view of $\mathbf{I}$ and $\mathbf{S}$—and

$$\mathsf{H}_0(\text{spin}) = A\mathbf{I} \cdot \mathbf{S}. \tag{2.43}$$

The solution to Eq. (2.43) is very simple and is identical to that of Eq. (2.19). If we use Eq. (2.41),

$$\epsilon_0(\text{spin}) = \frac{A}{2}[F(F+1) - I(I+1) - S(S+1)]. \tag{2.44}$$

Now with $I = \frac{1}{2}$ and $S = \frac{1}{2}$ there are two possible F values, $F = 1$ and $F = 0$. So we obtain from Eq. (2.44)

$$\epsilon_0(\text{spin}) = \begin{cases} \dfrac{A}{4} & \text{for} \quad F = 1 \\[2ex] \dfrac{-3A}{4} & \text{for} \quad F = 0 \end{cases}$$

The treatment of Eq. (2.43) is obviously identical to that of spin–orbit interaction for small λ values.

The magnetic interaction between a field and the electron and proton moments is given by Eq. (2.27) where

$$\boldsymbol{\mu} = -g_0\mu_B\mathbf{S} + g_I\mu_N\mathbf{I} \tag{2.45}$$

and a complete spin Hamiltonian is

$$\mathsf{H}(\text{spin}) = A\mathbf{I}\cdot\mathbf{S} + g_0\mu_B\mathbf{S}\cdot\mathbf{H} - g_I\mu_N\mathbf{I}\cdot\mathbf{H}. \tag{2.46}$$

We can determine two extreme types of solutions to Eq. (2.46). If we consider the vector-model diagram in Fig. 2.1*c* and Eq. (2.41), the solutions to Eq. (2.46) are similar to those of Eqs. (2.30) and (2.32) combined with Eq. (2.44). The result of following this algebra is that

$$\begin{aligned}\epsilon(\text{weak field}) = {}& \frac{A}{2}[F(F+1) - I(I+1) - S(S+1)] \\ & + g_0\mu_B H M_F \frac{F(F+1) + S(S+1) - I(I+1)}{2F(F+1)} \\ & - g_I\mu_N H M_F \frac{F(F+1) + I(I+1) - S(S+1)}{2F(F+1)}.\end{aligned} \tag{2.47}$$

This equation is not very useful. The field tends to rapidly uncouple **I** and **S**, because μ_B and μ_N differ by a factor of 1836. For fields of a 1000 gauss or so, Fig. 2.1*b* and Eq. (2.39) are fairly accurate, and the high field solution to Eq. (2.46) follows from Eq. (2.25) and (2.39), so that

$$\begin{aligned}\epsilon(\text{strong field}) = {}& AM_IM_S + g_0\mu_B HM_S \\ & - g_I\mu_N HM_I.\end{aligned} \tag{2.48}$$

This solution assumes that **I** and **S** are completely uncoupled by the field and that the Paschen–Back effect has taken place.

The states predicted by Eqs. (2.47) and (2.48) can be related by the fact that $M_F = M_S + M_I$. If we consider the states involving the maximum and minimum values of M_F, then $M_F\,(\text{max}) = 1 = \frac{1}{2} + \frac{1}{2}$ and $M_F(\text{min}) = -1 = -\frac{1}{2} - \frac{1}{2}$. Since these two F, M_F states can only be formed from unique M_S, M_I states, then Eqs. (2.39) and (2.41) must be identical for these two states. This can be demonstrated by comparing

Eqs. (2.47) and (2.48) for these two states and one obtains

$$\begin{aligned}\epsilon(\text{weak field})_{F=1,M_F=1} &= \frac{A}{4} + \frac{g_0\mu_B H}{2} - \frac{g_I\mu_N H}{2}\\ &= \epsilon(\text{strong field})_{M_I=M_S=1/2}\\ \epsilon(\text{weak field})_{F=1,M_F=-1} &= \frac{A}{4} - \frac{g_0\mu_B H}{2} + \frac{g_I\mu_N H}{2}\\ &= \epsilon(\text{strong field})_{M_I=M_S=-1/2}.\end{aligned} \tag{2.49}$$

Equation (2.49) must then be accurate for all fields for these two states. One can see for these two states that in Fig. 2.1*b* and 2.1*c* the energy is the same whether the angular momenta precess separately or together.

The two states for $M_F = 0 = M_I + M_S$ are not the same regardless of the coupling scheme. To solve for the energies at intermediate field a more complete quantum-mechanical calculation must be done. This can be done by using the following secular determinant,

$$\begin{vmatrix} \left(-\frac{A}{4} + \frac{x}{2}\right) - \epsilon & \frac{A}{2} \\ \frac{A}{2} & \left(-\frac{A}{4} - \frac{x}{2}\right) - \epsilon \end{vmatrix} = 0 \tag{2.50}$$

where $x = (g_0\mu_B + g_I\mu_N)H$. The terms along the diagonal of this determinant follow from Eq. (2.48). The $A/2$ term is the quantum-mechanical result, which tends to couple the momenta, and it arises from the part of the Hamiltonian $A(\mathsf{I}_X\mathsf{S}_X + \mathsf{I}_Y\mathsf{S}_Y)$. When this determinant is expanded and the resultant quadratic equation is solved, the result is

$$\epsilon = -\frac{A}{4} \pm \frac{1}{2}\sqrt{A^2 + x^2}. \tag{2.51}$$

In Fig. 2.2 the results of Eqs. (2.49) and (2.51) are plotted for the hydrogen atom. It can be seen from this figure that the Paschen–Back effect starts to become very effective at a few hundred gauss. It can also be seen, however, that although Eq. (2.48) is an excellent approximation to the $M_F = 0$ states above 1000 gauss, there is an appreciable shift between the actual levels and the full Paschen–Back approximation, even at rather high fields. An expansion of Eq. (2.51) for the case where $x \gg A$ gives a shift between the levels of Eqs. (2.48) and (2.51) which is

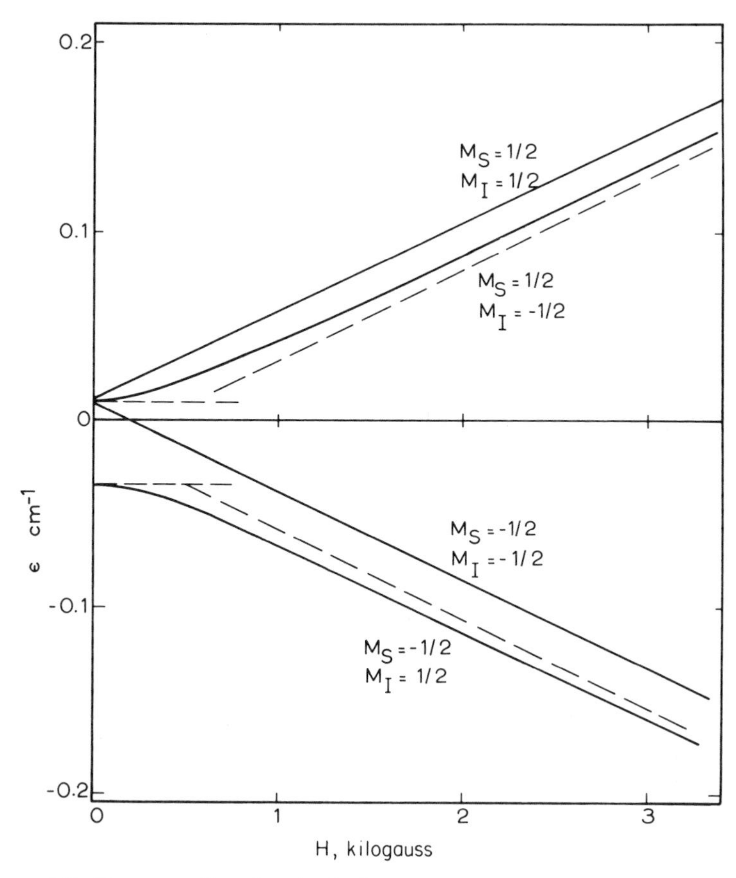

Figure 2.2. *Zeeman effect for the hydrogen atom. The dashed lines show the predictions of Eqs.* (2.47) *and* (2.48) *for* $M_F = 0$. *The field is shown in thousands of gauss.*

equal to $A^2/4x$. For the hydrogen atom this amounts to a shift of 6 cm^{-1} gauss. Thus a field of 10,000 gauss still has a shift from the Paschen–Back approximation of $6 \times 10^{-4}\ \text{cm}^{-1}$. This is the result of the small coupling of **I** and **S** that still remains.

In Chapter 8 we shall show that the energy levels in Fig. 2.2 can be measured with great accuracy by magnetic resonance spectroscopy. It is also interesting to note that these energy levels can be detected for hydrogen atoms in interstellar space by the direct observation of radio-frequency emission. It is even possible, with radio-frequency astronomy, to observe small magnetic field splittings for the hydrogen atoms in interstellar space.

References

E. U. Condon and G. H. Shortley, *The Theory of Atomic Spectra*, Cambridge University Press, New York (1953). The standard reference, very mathematical.

G. Herzberg, *Atomic Spectra and Atomic Structure*, Dover Publications, Inc., New York (1944). Not mathematical with emphasis on spectroscopy.

M. Karplus and R. N. Porter, *Atoms and Molecules: An Introduction for Students of Physical Chemistry*, W. A. Benjamin, Inc., Menlo Park, Calif. (1970). An introductory text that goes into detail.

C. E. Moore, *Atomic Energy Levels*, National Bureau of Standards Circular 467, Vols. 1 (1949), 2 (1952), and 3 (1958). The actual atomic levels as derived from atomic spectra.

L. Pauling and E. B. Wilson, Jr., *Introduction to Quantum Mechanics*, McGraw-Hill Book Company, New York (1935). A favorite older book on quantum mechanics including atomic structure.

J. C. Slater, *Quantum Theory of Atomic Structure*, 2 vols., McGraw-Hill Book Company, New York (1960). Quite complete, with a number of diagrams and examples.

Problems

2.1. The atomic structures of the first 20 elements in the periodic table can be explained by screened one-electron functions in which the order of increasing energy is $1s$, $2s$, $2p$, $3s$, $3p$, and $4s$. Use the symbolism $1s$, $(1s)^2$, $(1s)^2 2s$, etc., to represent the ground-state configurations of the atoms of the first 10 elements in the periodic table.

2.2. What is the ground-state-term symbol for H, He, Li, Be, and B according to the Pauli principle? (For B, neglect spin–orbit interaction.)

2.3. What are the three terms, without spin–orbit interaction, that are possible by the Pauli principle for the ground-state configuration of atomic carbon? Which has the lower energy according to Hund's rule about multiplicity?

2.4. What is the sign of λ for both B and C? Give their ground-state-term symbols, including spin–orbit interaction.

2.5.* There are also three possible terms for the ground-state configuration of atomic nitrogen, but the one with the highest multiplicity can be easily seen from the Pauli principle. What is its term symbol? How many unpaired electrons does it have? What is the value of g_J for this term?

2.6. What are g_J values for the two spin–orbit terms $^2P_{1/2}$ and $^2P_{3/2}$?

2.7. In the ions of the first-row transition metals it is found that the $3d$ orbitals seem to have a lower energy than do the $4s$. Under these conditions Ti^{+3} and V^{+3} have the closed-shell configuration of Ar but with the additional one and two electrons, respectively, in $3d$ orbitals. What is the ground-state term for Ti^{+3}, neglecting spin–orbit interaction? How many possible terms are there for V^{+3}? The lowest energy term is the one of highest multiplicity with the higher orbital angular momentum. (The answer to this question will be found in Table 3.1.)

2.8.* Express as a power series in x the difference in energies of the hydrogen-atom levels with the same M_I value but with different M_S values. What values of energy do the two differences approach as x becomes very large and when $M_I = \pm\frac{1}{2}$?

2.9. Hydrogen atoms in interstellar space can be detected by zero-field emission of energy. Use the relation $\Delta\epsilon = h\nu$ to calculate the zero-field absorption frequency of hydrogen atoms in megahertz. What is the vacuum wavelength of this radiation in centimeters?

Solutions:

2.5. The ground configuration of nitrogen atom is $(1s)^2(2s)^2(2p)^3$. If we were to prepare a list like Table 2.2 for this configuration, the Pauli principle would only allow permutations of the quantum numbers for the $(2p)^3$ electrons. The highest possible M_S value for three electrons is $\frac{3}{2}$, and this will give a multiplicity for the lowest energy term of 4. At the same time, if the three electrons all have the same value for $M_S(i)$, they must have different values for $M_l(i)$. For $l = 1$ there would be only one possibility and this will give $M_L = 0$. Although we have only discussed one line in Table 2.2 for $(2p)^3$, we can clearly see that the ground-state term must be ^4S. To obtain g_J we use the fact that $J = S$, and this gives $g_J = 2$.

2.8. If we start with Eq. (2.51) for $A/x \ll 1$ and $M_F = 0$,

$$\begin{aligned}\epsilon &= -\frac{A}{4} \pm \frac{x}{2}\sqrt{1 + \left(\frac{A}{x}\right)^2} \\ &= -\frac{A}{4} \pm \frac{x}{2}\left(1 + \frac{A^2}{2x^2} + \cdots\right) \\ &= -\frac{A}{4} \pm \frac{x}{2} \pm \frac{A^2}{4x}.\end{aligned}$$

Comparison with Eq. (2.48) shows that the positive sign must correspond to

$M_S = \frac{1}{2}$, $M_I = -\frac{1}{2}$, and the negative sign to $M_S = -\frac{1}{2}$, $M_I = \frac{1}{2}$.

If we now subtract these results from those of Eq. (2.49) for the same M_I value but different M_S values, we obtain

$$\epsilon_{1/2} - \epsilon_{-1/2} = g_0\mu_B \pm \frac{A}{2} + \frac{A^2}{4x}$$

where the top sign is for $M_I = \frac{1}{2}$ and the bottom is for $M_I = -\frac{1}{2}$. This result is similar to what one would obtain from Eq. (2.48) except for the last term, which corrects for the deviation from the Paschen–Back limit. (*Note:* This energy difference is what is observed in EPR.)

3

Atomic Paramagnetism and the Effects of Crystalline Fields

THE ENERGY RELATIONSHIPS of Chapter 2 can be combined with the basic equations of Chapter 1 to yield theoretical susceptibilities for bulk samples of atoms or ions. We will find that these predictions are quite valid for gases, but for ions in solution or in hydrated crystals these equations fail. The reason for this is that in condensed matter the atoms are always subjected to the electric fields of neighboring atoms. These electric fields have a great effect upon the orbital motion of the electrons. The simplest theory for the treatment of this interaction is contained in what is called "crystal field theory." This theory can be widely applied and we shall illustrate its use for the first-row transition-metal ions. Although this theory was originally developed to explain the magnetic properties of the transition-metal ions, it can also be used to account for the stability of their complexes and to explain a number of their physical properties. This theory also serves as a good introduction to the magnetic properties of molecules. But before we consider crystal field theory, let us use the results of Chapters 1 and 2 to predict the susceptibilities of free atoms.

3.1. Susceptibilities of Free Atoms

J = S CASE

The simplest case is that of a sample of atoms where each is in an S state so that $L = 0$ and there is no orbital angular momentum. To simplify it further assume a doublet ground state with $S = \frac{1}{2}$ and with all excited electronic states much greater than kT in energy. This would correspond to a sample of hydrogen atoms in $^2S_{1/2}$ states, if we also neglect its hyperfine interaction.

From Fig. 2.2, if we neglect the hyperfine effects, or from Eq. (2.29) with $\mathbf{J} = \mathbf{S}$ and with $g_J = g_0$, one can see in Eq. (1.20) that

$$\epsilon^{(1)}H = g_0\mu_B H M_S \tag{3.1a}$$

$$\epsilon^{(2)}H \approx 0 \tag{3.1b}$$

where $M_S = \pm\frac{1}{2}$. We could simply use Eq. (1.27), but for this example we shall derive our result using the Boltzmann relation. This relation states that at thermal equilibrium there will be more atoms with $M_S = -\frac{1}{2}$ than there will be with $M_S = \frac{1}{2}$. The quantitative result is

$$\frac{N_{-1/2}}{N_{1/2}} = e^{\Delta\epsilon/kT} = e^{g_0\mu_B H/kT}, \tag{3.2a}$$

and if $g_0\mu_B H \ll kT$,

$$\frac{N_{-1/2}}{N_{1/2}} = 1 + \frac{g_0\mu_B H}{kT} \tag{3.2b}$$

and, in a convenient form,

$$N_{-1/2} - N_{1/2} = \left[N_{-1/2} + N_{1/2}\right]\frac{g_0\mu_B H}{2kT} = N\frac{g_0\mu_B H}{2kT}. \tag{3.2c}$$

Eq. (3.2c) is quite valid at room temperature where $kT \approx 200\ \text{cm}^{-1}$ and, for ordinary magnetic fields, $\mu_B H \approx 1\ \text{cm}^{-1}$.

The two values of μ_Z are

$$\mu_{Z(-1/2)} = -\mu_{Z(1/2)} = \frac{g_0\mu_B}{2}, \tag{3.3}$$

and from Eq. (1.23),

$$\chi_M = (N_{-1/2} - N_{1/2})\frac{g_0\mu_B}{2H}, \tag{3.4}$$

so that

$$\chi_M = N\frac{g_0^2\mu_B^2}{4kT}. \tag{3.5}$$

In deriving this result we have developed the answer in steps rather than substituting Eq. (3.1) into Eq. (1.27) and summing over the two levels. This is to emphasize that χ_M depends upon the difference between the populations of the two levels induced by the field.

This same result can be obtained from Eq. (1.16). For this spin-only situation, since $\mathbf{S}^2$ is quantized in LS coupling,

$$\boldsymbol{\mu}^2 = g_0^2\mu_B^2 S(S+1), \tag{3.6a}$$

and for $S = \frac{1}{2}$,

$$\boldsymbol{\mu}^2 = \frac{3g_0^2\mu_B^2}{4}. \tag{3.6b}$$

When this is substituted in Eq. (1.16), the result is identical with Eq. (3.5).

The value of μ_{eff} can be obtained from Eq. (3.6*a*) with the result that

$$\mu_{\text{eff}} = g_0\sqrt{S(S+1)}. \tag{3.7}$$

For our example, $S = \frac{1}{2}$ and μ_{eff} is very close to $\sqrt{3}$. When an atom has only a spin magnetic moment one finds that Eq. (1.16) and the value of μ_{eff} given by Eq. (3.7) will nicely fit the observed susceptibility. It is common practice to consider an agreement between an experimental value for μ_{eff} and the possible values predicted by Eq. (3.7) as confirmation of a "spin-only" magnetic moment. In Fig. 1.5, data were shown for gaseous O_2 where μ_{eff} was very close to $\sqrt{8}$ or $2\sqrt{2}$. Oxygen is an excellent example of a complex species possessing a spin-only moment consistent with $S = 1$. The reasons for this will be discussed in Chapter 4.

If the hyperfine interaction is included for hydrogen atoms we have the four energy levels given by Eqs. (2.49) and (2.51). At high or low fields these can be used to determine $\epsilon^{(1)}$ and $\epsilon^{(2)}$ values suitable for use in Eq. (1.27). Since all four energy levels are closely spaced with respect to kT, they all are "low-frequency" terms as defined by Van Vleck. This means that both terms in Eq. (1.27) will combine to yield an answer equivalent to the first term in Eq. (1.16) and one will find that hyperfine splittings do not affect χ_M at ordinary temperature. We shall illustrate the use of the "low-frequency" concept for atoms with orbital angular momentum.

One might want to calculate α for ${}^2S_{1/2}$ atoms and this involves a determination of $\epsilon^{(2)}$. Because of spin–orbit interaction neither L_Z nor S_Z is exactly quantized, but the sum $\mathsf{L}_Z + \mathsf{S}_Z = \mathsf{J}_Z$ is quantized. For convenience, the Zeeman Hamiltonian in Eq. (1.19) can be arranged as

$$\mathsf{H}'(\text{para}) = -\mu_B \mathsf{J}_Z H - \mu_B(g_0 - 1)\mathsf{S}_Z H. \tag{3.8}$$

Since J_Z is quantized in field-free atoms, the first term in Eq. (3.8) contributes only to $\epsilon^{(1)}$ and not at all to $\epsilon^{(2)}$ in Eq. (1.30), so that

$$\epsilon_i^{(2)} = -\mu_B^2(g_0 - 1)^2 \sum_{j \neq i} \frac{|\langle\psi_i^0|\, \mathsf{S}_Z\, |\psi_j^0\rangle|^2}{\epsilon_j^0 - \epsilon_i^0}\,. \tag{3.9}$$

This is nothing but a more convenient form for Eq. (1.30), suitable only for atoms. We shall make quantitative use of this equation in Section 3.2, but for the hydrogen atom it suffices to say that this will give it a very small value for the paramagnetic contribution to α.

J = L + S CASE

If the ground state is a ${}^2P$, ${}^3P$, ${}^2D$, etc., we must include the orbital moment. Since a magnetic field will tend to uncouple **L** and **S** we have to consider the Paschen–Back effect, but for spin–orbit splittings greater than 1 cm^{-1} we can start with the assumption of coupled **L** and **S**. In this case, if we neglect $\epsilon^{(2)}$ for the moment,

$$\epsilon^{(1)} = g_J \mu_B M_J, \tag{3.10}$$

where we have used Eq. (2.30). If we then sum Eq. (1.27) over the

$2J + 1$ values of M_J, we find

$$\chi_M = \frac{N g_J^2 \mu_B^2}{(2J+1)kT} \sum_{M_J} M_J^2, \tag{3.11}$$

and if we use an important identity,

$$\sum_{M_J} M_J^2 = \tfrac{1}{3} J(J+1)(2J+1), \tag{3.12}$$

then

$$\chi_M = \frac{N g_J^2 \mu_B^2}{3kT} J(J+1). \tag{3.13}$$

It is clear that this equation is in agreement with Eq. (1.16) and the definition of g_J. We now have

$$\mu_{\text{eff}} = g_J \sqrt{J(J+1)}, \tag{3.14}$$

where we have neglected $\epsilon^{(2)}$ and have assumed that only the lowest spin–orbit term is thermally populated.

The calculation of $\epsilon^{(2)}$ involves a good deal of quantum mechanics for the use of Eq. (3.9). The matrix elements for S_Z in the representation where $\mathbf{S}^2$, $\mathbf{L}^2$, and $\mathbf{J}^2$ are quantized are tabulated in books such as Condon and Shortley. Instead of the general case let us consider a ground ^2P state and its two $^2P_{1/2}$ and $^2P_{3/2}$ terms. For these two, Eq. (2.20*b*) tells us that the $^2P_{1/2}$ is lower in energy by $3\lambda/2$. Since we must sum Eq. (3.9) over the M_J values we can use the "sum rules," and from tabulated matrix elements one finds that

$$\sum_{M_J} |\langle {}^2P_{1/2}| \, \mathsf{S}_Z \, |{}^2P_{3/2}\rangle|^2 = 4/9. \tag{3.15}$$

So if we only consider the two ^2P terms from Eqs. (1.27) and (3.9),

$$\alpha(\text{para}) = \frac{2N\mu_B^2(g_0 - 1)^2}{2J+1} \, \frac{4/9}{3\lambda/2} \tag{3.16}$$

where for the $^2P_{1/2}$ ground state $J = \frac{1}{2}$. When we combine Eqs. (3.13) and (3.16), we have the complete χ_M if the $^2P_{3/2}$ is not thermally populated.

If $kT \approx \lambda$, the $^2P_{3/2}$ will also be populated in Eq. (1.27). The complete equation can be readily found and we have $g_J = \frac{2}{3}$ and $\frac{4}{3}$ for the $^2P_{1/2}$ and $^2P_{3/2}$ states, respectively. Then, if we neglect diamagnetism,

$$\chi_M = \frac{N\mu_B^2}{3kT} \frac{\frac{2}{3} + \frac{80}{3}e^{-x}}{2 + 4e^{-x}} + \frac{N\mu_B^2(g_0 - 1)^2 \frac{8}{9}(1 - e^{-x})}{xkT(2 + 4e^{-x})} \tag{3.17}$$

where $x = 3\lambda/2kT$. This complete equation is rather complicated.

For the special circumstance that $kT \gg \lambda$, then $x \ll 1$, and we can expand the exponentials in Eq. (3.17). The first nonvanishing term follows the Curie law with

$$\chi_M = \frac{N\mu_B^2}{3kT}\left[\frac{\frac{82}{3}}{6} + \frac{(g_0 - 1)^2 \frac{8}{9}}{2}\right], \tag{3.18a}$$

and, if $g_0 = 2.0$, we have

$$\chi_M = \frac{5N\mu_B^2}{3kT}. \tag{3.18b}$$

This result can be summarized with the observation that $\mu_{\text{eff}} = \sqrt{5}$ is the high-temperature limit for a ^2P term.

Such a simple result is explained by the general validity of Eq. (1.16). When $kT \gg \lambda$, both terms in Eq. (1.27) contribute "low-frequency" elements that form the average moment in Eq. (1.31*b*). The problem is only that of establishing the correct average value for $\boldsymbol{\mu}^2$. One method for this is to realize that when $kT \gg \lambda$, the coupling between **L** and **S** is essentially averaged to zero. If we were to ignore any coupling between **L** and **S**, we could assume that they precess separately. In this case we should have the result that $(\boldsymbol{\mu}^2)_{\text{av}}$ is the sum of squares of the orbital and spin moments with

$$(\boldsymbol{\mu}^2)_{\text{av}} = \mu_B^2[L(L + 1) + g_0^2S(S + 1)]$$

or

$$\mu_{\text{eff}} = \sqrt{L(L + 1) + g_0^2S(S + 1)}. \tag{3.19a}$$

With $L = 1$ and $S = \frac{1}{2}$ this gives $\mu_{\text{eff}} = \sqrt{5}$, in exact agreement with the rigorous high-temperature limit of Eq. (3.18*b*).

In some ways it does not seem accurate to say that **L** and **S** precess separately at high temperatures. If both the $J = \frac{1}{2}$ and $J = \frac{3}{2}$ terms are populated, it is not necessarily true that **L** and **S** are uncoupled. This is illustrated if we form the total moment as the vector sum with

$$(\boldsymbol{\mu}^2)_{av} = \mu_B^2(\mathbf{L} + g_0\mathbf{S})^2_{av}$$
$$= \mu_B^2[L(L+1) + g_0^2 S(S+1)]$$
$$+ 2g_0\mu_B^2(\mathbf{L}\cdot\mathbf{S})_{av}. \tag{3.19b}$$

To obtain Eq. (3.19*a*) it is only necessary that $(\mathbf{L}\cdot\mathbf{S})_{av} = 0$. If we assume, as we did before, that **L** and **S** are uncoupled, $\mathbf{L}\cdot\mathbf{S}$ is equal to $M_L M_S$. This averages to zero as we sum over all the M_L and M_S values. If we assume that **L** and **S** are coupled, $\mathbf{L}\cdot\mathbf{S}$ is given by Eq. (2.19). It is not too difficult to show that this also averages to zero when the $2J + 1$ degeneracy is included in the averaging process over all the J values. As a result, we can see that Eq. (3.19*a*) is the correct high-temperature limit whether or not we assume that **L** and **S** are still coupled or not. The key point is that $\mathbf{L}\cdot\mathbf{S}$ averages to zero over an entire spin–orbit multiplet.

The vector model can also be used to explain the matrix element in Eq. (3.15) if we use Eq. (1.33). In this equation if there is a single excited state with energy $\Delta\epsilon$, then by matrix multiplication, if we again neglect diamagnetism,

$$\alpha = \frac{2}{3}\frac{\langle\psi_0^0|\,\boldsymbol{\mu}^2\,|\psi_0^0\rangle}{\Delta\epsilon}. \tag{3.20}$$

The magnetic moment in Eq. (3.20) is not the magnetic moment in Eqs. (3.19) or (3.14), but it can be found in the vector-model diagram in Fig. 3.1. The magnetic moment in Eq. (3.14) is parallel to J, but as we can see from Fig. 3.1 there is a component of the moment perpendicular to **J**. From Eq. (3.14) we know that

$$\mu_{\|} = -g_J\mu_B\sqrt{J(J+1)}, \tag{3.21}$$

and from Fig. 3.1 we can see that

$$\mu_\perp^2 = (g_0 - 1)^2\mu_B^2 S(S+1) - (g_J - 1)^2\mu_B^2 J(J+1). \tag{3.22}$$

With $g_0 = 2.0$, $g_J = \frac{2}{3}$, $J = \frac{1}{2}$, and $\Delta\epsilon = 3\lambda/2$, Eq. (3.22) supplies a moment in Eq. (3.20) which will give the same result as Eq. (3.16). We can see that the precession of **L** and **S** around **J** in the vector model does not allow $\mu_\perp$ to contribute to $\epsilon^{(1)}$, but this moment forms the classical

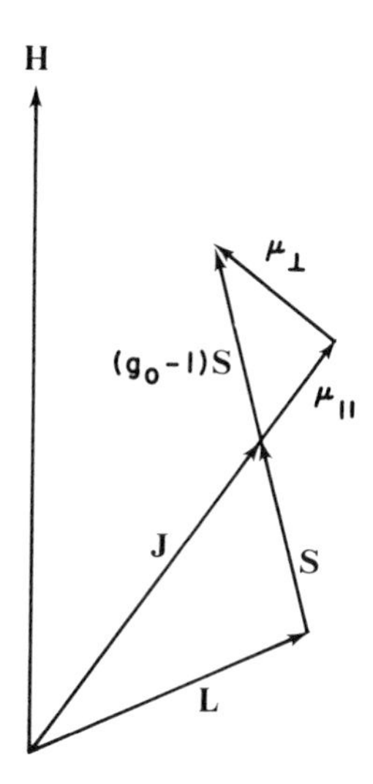

Figure 3.1. *Parallel and perpendicular components of the magnetic moment formed by the equation*

$$\boldsymbol{\mu} = -\mu_B(\mathbf{L} + g_0\mathbf{S}) = -\mu_B[\mathbf{J} + (g_0 - 1)\mathbf{S}].$$

(Note: $\boldsymbol{\mu}$ *is shown pointing in the same direction as* **J** *for clarity. It should be remembered that* **J** *precesses about* H*, and* **S** *and* **L** *precesses about* **J**.)

basis for $\epsilon^{(2)}$. We will find other systems where $\epsilon^{(2)}$ is due to the presence of a magnetic moment which can be included in the vector model.

$kT \ll \mu_B H$ CASE

For very low temperatures the expansion of the exponentials in Eq. (1.26) must include more terms. In the extreme case where $kT \ll \mu_B H$, only the lowest magnetic sublevel is populated and Eq. (1.27) has no validity. For this circumstance we must go back to Eq. (1.23) and M_Z becomes a constant that is more or less independent of both temperature and field. The magnetization is saturated and

$$M_Z(\text{sat.}) = N\mu_B g_J J \tag{3.23}$$

where $(-J)$, in this equation, is the minimum value of M_J. In Fig. 3.2 a plot is given of $M_Z/N\mu_B$ vs. H/T. For Fe^{+3} the saturation value is in agreement with a ${}^6S_{5/2}$ ion where $J = \frac{5}{2}$ and $g_J = 2$. However, the $Fe(CN)_6^{-3}$ ion acts as though in the complex only one of the electrons is unpaired. The lines drawn in Fig. 3.2 were determined using Eqs. (1.23)

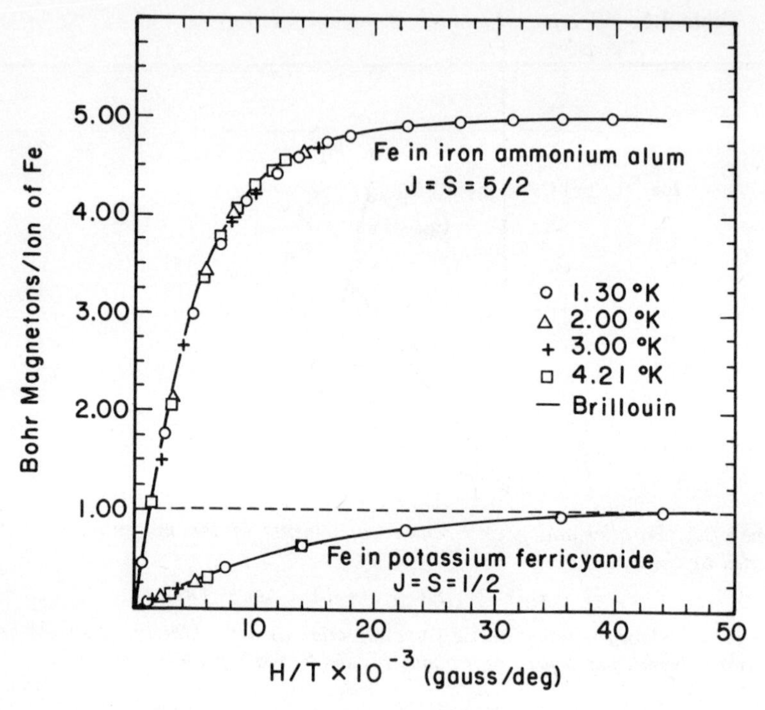

Figure 3.2. *Magnetization of* Fe^{+3} *and* $Fe(CN)_6^{-3}$ *approaching saturation at high fields and low temperatures.* [*Data from W. E. Henry, Phys. Rev.*, **88,** 559 (1952).]

and (1.25) and these data are an excellent confirmation of quantum theory. Classical theory would predict an M_Z (sat.) value with J in Eq. (3.23) replaced by $\sqrt{J(J+1)}$.

EFFECTS OF CRYSTALLINE FIELDS

While Fe^{+3} follows free ion theory very nicely, most of the other transition-metal ions do not. In Table 3.1 the μ_{eff} values for the iron group and lanthanide ions are given for aqueous solution. These values were derived from susceptibility data in dilute solution and μ_{eff} was determined by Eq. (1.17). It is seen that the iron-group ions do not follow Eq. (3.14) but are fit rather well by a spin-only approximation using Eq. (3.7); on the other hand, the lanthanide ions do follow Eq. (3.14) fairly well. The reasons for this difference have to be discussed in terms of the interactions between these ions and the surrounding water molecules in solution.

Table 3.1. *Effective Magnetic Moments for Ions in Aqueous Solution*

		Free ion			μ_{eff}		
n	Ion	term	Spin–orbit[a]	g_J	observed[b]	$g_J\sqrt{J(J+1)}$	$2\sqrt{S(S+1)}$
			Iron group: $(3d)^n$				
1	V^{+4}	$^2D_{3/2}$	248	4/5	1.73[c]	1.55	1.73
2	V^{+3}	$^3F_2$	105	3/2	2.78	1.63	2.83
3	Cr^{+3}	$^4F_{3/2}$	87	2/5	3.78	0.77	3.87
4	Cr^{+2}	$^5D_0$	57	—	4.80	0.00	4.90
5	Mn^{+2}	$^6S_{5/2}$	—	2	5.91	5.92	5.92
6	Fe^{+2}	$^5D_4$	−100	3/2	5.33	6.70	4.90
7	Co^{+2}	$^4F_{9/4}$	−180	4/3	5.04[d]	6.63	3.87
8	Ni^{+2}	$^3F_4$	−335	5/4	3.22	5.59	2.83
9	Cu^{+2}	$^2D_{5/2}$	−852	6/5	1.94	3.55	1.73
			Lanthanide group: $(4f)^n$				
1	Ce^{+3}	$^2F_{5/2}$	640	6/7	2.50[d]	2.54	1.73
2	Pr^{+3}	$^3H_4$	800	4/5	3.4[d]	3.58	2.83
3	Nd^{+3}	$^4I_{9/2}$	900	8/11	3.45[d]	3.62	3.87
4	Pm^{+3}	$^5I_4$	(1070)	3/5	—	2.68	4.90
5	Sm^{+3}	$^6H_{5/2}$	1200	2/7	1.63[d]	0.84	5.92
6	Eu^{+3}	$^7F_0$	1410	—	3.96[d]	0.00	6.93
7	Gd^{+3}	$^6S_{7/2}$	1540	2	7.86	7.94	7.94
8	Tb^{+3}	$^7F_6$	1770	3/2	9.76	9.72	6.93
9	Dy^{+3}	$^6H_{15/2}$	1860	4/3	10.85	10.63	6.93
10	Ho^{+3}	$^5I_8$	2000	5/4	10.39	10.60	4.90
11	Er^{+3}	$^4I_{15/2}$	2350	6/5	9.54	9.59	3.87
12	Tm^{+3}	$^3H_6$	2660	7/6	—	7.57	2.83
13	Yb^{+3}	$^2F_{7/2}$	2940	8/7	4.50	4.54	1.73

[a] Values in cm^{-1}. For the iron group the LS coupling constant λ is given; for the lanthanide group the one-electron constant $\zeta(n, l)$ is given. The lanthanide values are taken from B. Bleaney, *Proc. Phys. Soc.* (*London*), **A68,** 937 (1955).
[b] Taken from the same source as Table 1.2.
[c] The V^{+4} value is for VO^{+2}. The other ions are probably not appreciably hydrolyzed.
[d] For these ions tabulated data indicate that they follow a Curie–Weiss law with an appreciable Δ value.

3.2. Crystal Field Theory and Paramagnetism

In the atomic Hamiltonian, Eq. (2.4), it was assumed that the atom or ion was not perturbed by its surroundings and that its electrons were free to move in a more-or-less spherical potential function. When an ion is placed in a solid or in solution, this is no longer true. The other ions in the solid, or the coordinated water molecules in solution, perturb the electrons of an ion, so that they must move in a different way. The most elementary way of including this effect is to add another potential-energy term to Eq. (2.4). This method is known as the crystal field model and it has been very satisfactory for taking into account many of the effects of the environment upon the energy levels of ions. It was first formulated as a theory by H. Bethe (1929) and as a tool for paramagnetism by Van Vleck (1932). A more complete treatment must include the energy levels of the surrounding molecules into a general molecular-orbital treatment. This more complete treatment is often called ligand field theory and this more general name can be used to designate all levels of approximation. The crystal field approximation is quite satisfactory for most purposes and like many techniques in science its usefulness depends a great deal upon its simplicity and generality. A short discussion of the molecular-orbital treatment is given in Chapter 4.

The crystal field model simply adds a potential-energy term to Eq. (2.4), so that

$$\mathsf{H}_0(\text{atom}) = \mathsf{H}(\text{atom}) + \mathsf{V}(\text{crystal field}). \tag{3.24}$$

This potential-energy term has two important aspects. First, how much energy does it contribute? And second, what is its symmetry? The energy of the crystal field term has to be considered in relation to the other terms in Eq. (2.4). If the crystal field term is less important than is spin–orbit interaction, it would be best to include this element into the calculation after **J** is formed from **L** + **S**. But if the crystal field term is more important than is spin–orbit interaction, **J** may have no particular significance. If, in addition, the crystal field energy is larger than the electron–electron repulsion effects, then our simple atomic wave functions might be a poor approximation and it should be necessary to use the more complete ligand field theory. On the basis of experience the following classification can be made:

1. Low-field case—lanthanide ions (4f): crystal field energy $<$ spin–orbit interaction.
2. Medium† (weak)-field case—transition ions (3d): crystal field energy $>$ spin–orbit interaction.
3. High† (strong)-field case—transition ions (4*d*) and (5*d*) (also covalent complexes of 3*d* ions): crystal field energy $\approx$ electron–electron repulsion.

The symmetry of the crystal field is very important. In fact, most of the

† These are more commonly called the weak- and strong-field cases, respectively, and we shall follow this custom.

results of this theory are a direct outgrowth of the fact that it is basically a theory based upon symmetry. In general terms, the free atom's wave functions are determined by its basic spherical symmetry, and crystal field theory examines the effects of the decrease in this symmetry by the surroundings. In many solids each ion is surrounded by six or eight nearest neighbors arranged in a regular array. In aqueous solution the cations are always hydrated and they are often surrounded by six close water molecules. A study of Cr^{+3}, Co^{+3}, and other ions in complexes has shown that these ions usually have an octahedral coordination, and

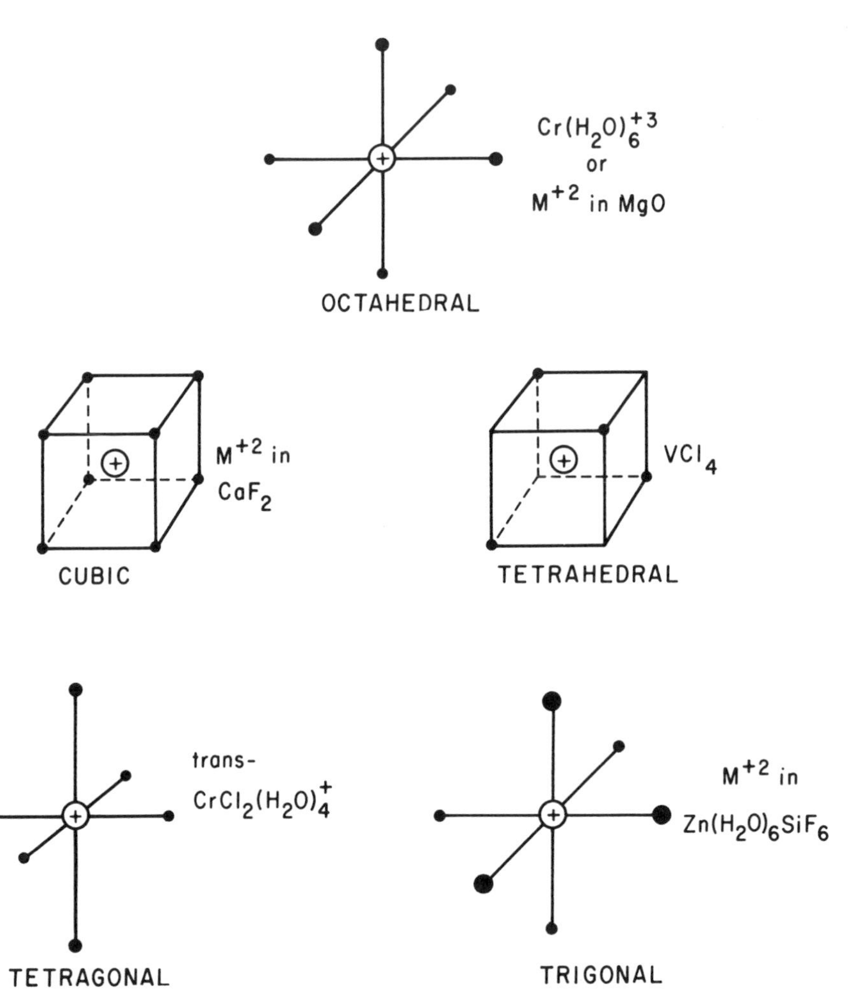

Figure 3.3. *Some important crystal fields. The heavy black dots represent negatively charged groups.*

crystal field theory uses this symmetry as a starting point. In Fig 3.3 we give some common coordination symmetries and examples.

TRANSITION-METAL IONS (3*d*)

The iron-group transition-metal ions form many complexes and their properties can be explained either by a weak field, or for covalent types, by strong crystal fields. These ions have a $(3d)^n$ configuration and the lowest state configurations of several of these ions is given in Table 3.2.

Table 3.2. *Effects of Octahedral Crystal Fields upon the Iron-Group Ions*

			Lowest energy levels[a]			
			Octahedral crystal field			
			weak field		strong field	
Configuration	**Example**	**Free ion**	configuration	term	configuration	term
$(3d)^1$	Ti^{+3}	^2D	t_{2g}	$^2T_{2g}$	t_{2g}	$^2T_{2g}$
$(3d)^2$	V^{+3}	^3F	$(t_{2g})^{1.8}(e_g)^{0.2}$	$^3T_{1g}$	$(t_{2g})^2$	$^3T_{1g}$
$(3d)^3$	Cr^{+3}	^4F	$(t_{2g})^3$	$^4A_{2g}$	$(t_{2g})^3$	$^4A_{2g}$
$(3d)^4$	Cr^{+2}	^5D	$(t_{2g})^3e_g$	^5E_g	$(t_{2g})^4$	$^3T_{1g}$
$(3d)^5$	Mn^{+2}	^6S	$(t_{2g})^3(e_g)^2$	$^6A_{1g}$	$(t_{2g})^5$	$^2T_{2g}$
$(3d)^6$	Fe^{+2}	^5D	$(t_{2g})^4(e_g)^2$	$^5T_{2g}$	$(t_{2g})^6$	$^1A_{2g}$
$(3d)^7$	Co^{+2}	^4F	$(t_{2g})^{4.8}(e_g)^{2.2}$	$^4T_{1g}$	$(t_{2g})^6e_g$	^2E_g
$(3d)^8$	Ni^{+2}	^3F	$(t_{2g})^6(e_g)^2$	$^3A_{2g}$	$(t_{2g})^6(e_g)^2$	$^3A_{2g}$
$(3d)^9$	Cu^{+2}	^2D	$(t_{2g})^6(e_g)^3$	^2E_g	$(t_{2g})^6(e_g)^3$	^2E_g

[a] Without spin–orbit interaction.

The first effect of V(C.F.) in Eq. (3.24) is to raise the energy of all the electrons in a cation, since the coordinating groups are negative and they repel the electrons of the central ions. This energy would contribute to the lattice energy of an ion and it could modify the radial part of the wave function. The second effect of V(C.F.) is to interact strongly with the orbital motion of the electrons. When the electrons try to rotate against the crystal fields, these fields will deflect the electrons, so that they are no longer rotating freely. In mathematical terms, we can say that V(C.F.) does not commute with either $\mathbf{L}^2$ or L_Z. The coordinates that define orbital angular momentum also define the crystal field, and momenta and conjugate coordinates do not commute.

The starting wave functions for a crystal field calculation should express the fact that L_z is not quantized.† For the weak-field case we can still use one-electron wave functions that are eigenfunctions of $\mathbf{L}^2(i)$ but not of $L_z(i)$. This is very similar to the free ion with the introduction of spin–orbit interaction. The most common 3*d* wave functions that are used in this case are shown in Fig. 3.4. These functions are linear combinations of the 3*d* eigenfunctions of $L_z(i)$ and they are real positive functions with maximum electron densities along a convenient coordinate system. It can be seen from a comparison of Figs. 3.3 and 3.4 that the

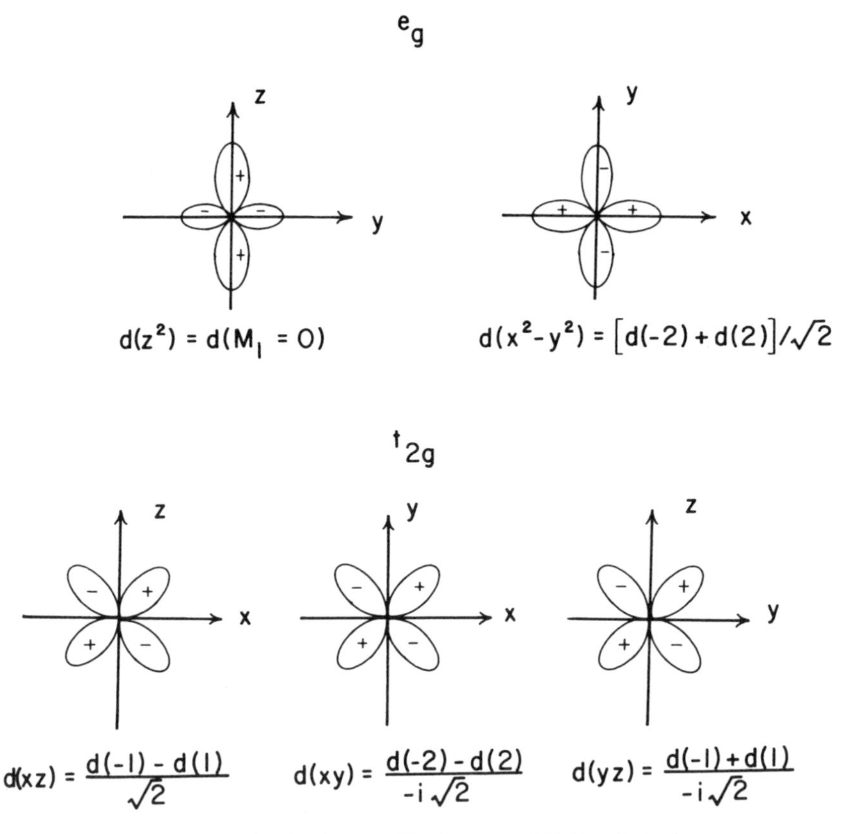

Figure 3.4. *d orbitals suitable for crystal field calculations.*

† We will introduce the coordinates *x*, *y*, and *z* to designate the crystal field coordinate system and reserve *X*, *Y*, and *Z* to designate the laboratory fixed-coordinate system with *Z* as the magnetic field axis.

$d(xz)$, $d(yz)$, and $d(xy)$ functions have their maximum electron densities pointed away from the negative charges in octahedral coordination, whereas the $d(z^2)$ and $d(x^2 - y^2)$ point directly at these charges.

The result of this difference is that for octahedral fields, V(C.F.) splits the 3*d* orbitals into two orbitals with different energies. This splitting is shown in Fig. 3.5*a*. The lower energy orbital is called the t_{2g} and it is composed of the $d(xz)$, $d(yz)$, and $d(xy)$. The higher energy orbital is

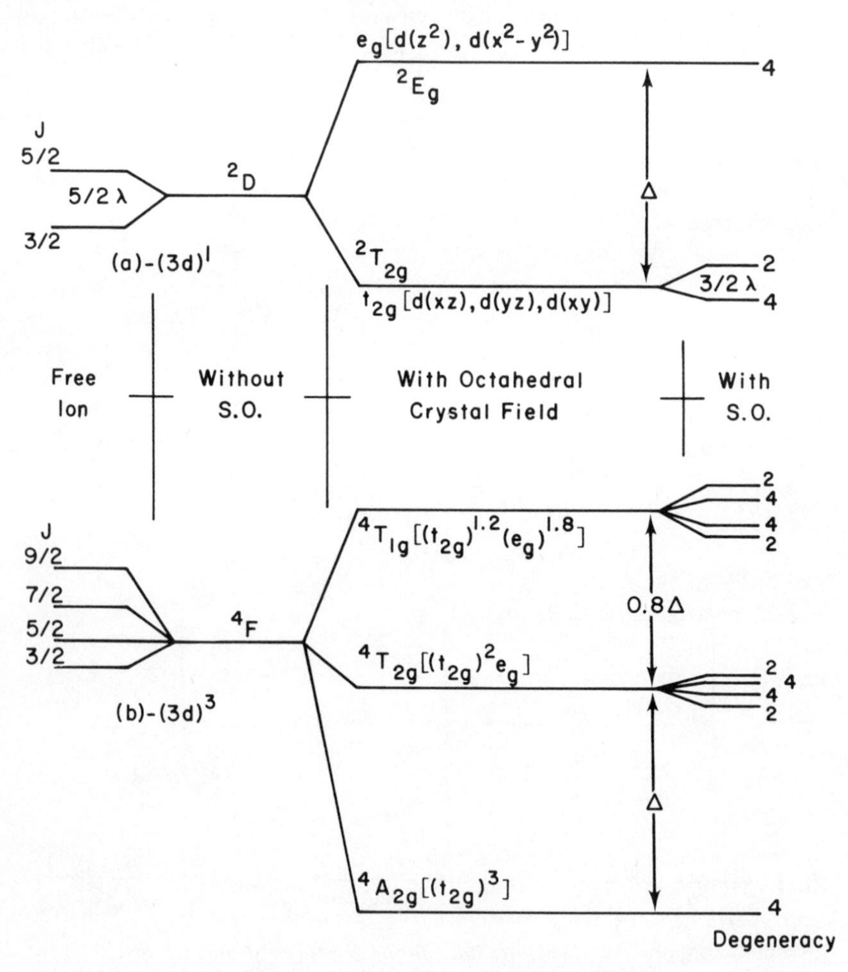

Figure 3.5. *Effect of a weak octahedral crystal field on (a) the* $(3d)^1$ *and (b) the* $(3d)^3$ *configurations. (Note: Not drawn to scale for* $\Delta \gg \lambda \approx 100$ *cm*$^{-1}$*. The value of* Δ *for the* $Ti(H_2O)_6^{+3}$ *ion is found to be* 20,000 *cm*$^{-1}$*, for the* $^2T_{2g} \rightarrow {}^2E_g$ *transition is at* 5000 Å *for* Ti^{+3} *in aqueous solution.)*

called the e_g and it is composed of the $d(z^2)$ and $d(x^2 - y^2)$. In an octahedral complex-ion group theoretical, or symmetry, considerations show that the fivefold orbital degeneracy of a $3d$ electron is broken into a threefold ($t = 3$) and a twofold ($e = 2$). The symbols t_{2g} and e_g are the accepted group-theoretical nomenclature for octahedral symmetry.

The splitting Δ is a measure of the strength of crystal field. For weak crystal fields Δ should be smaller than the splitting between terms but larger than the spin–orbit interaction. Group-theoretical designations replace the S, P, D, etc., in the term symbols, and the term symbol for the lowest state of Ti^{+3} $(3d)^1$ in an octahedral crystal field is $^2T_{2g}$. The first excited term is ^2E_g as shown in Fig. 3.5*a*. All this neglects spin–orbit interaction, which will be considered later.

If more than one $3d$ electron is present, the splitting pattern is more complicated. For weak crystal fields we need only consider the splitting of the lowest free-ion term. These terms are shown for $(3d)^n$ in Table 3.2; ^2D terms are lowest for $n = 1$ or 9, and ^5D are lowest for $n = 4$ or 6. Group theory shows that in the presence of an octahedral crystal field, D terms are split according to the rule

$$D \rightarrow T_{2g} + E_g. \tag{3.25a}$$

The lower energy is the T_{2g} for $n = 1$ or 6 but it is the E_g for $n = 4$ or 9. The F terms follow the rule

$$F \rightarrow T_{1g} + T_{2g} + A_{2g}. \tag{3.25b}$$

The order of the energies is more complicated for F terms, but for the highest possible multiplicities it is $T_{1g} < T_{2g} < A_{2g}$ for $n = 2$ or 7 and just the opposite order for $n = 3$ or 8. The ^6S term for $(3d)^5$ is unsplit, since it has no orbital degeneracy. Its weak-field-term symbol is $^6A_{1g}$ where A means no orbital degeneracy.

LOW-SPIN COMPLEXES

In strong crystal fields the term with the lowest energy may be different from that in weak crystal fields. One can see in Table 3.2 that strong fields pair some of the electrons in the ground-state terms for $n = 4$ through $n = 7$. This can be most easily understood if we use the t_{2g} and e_g orbitals as one-electron wave functions. For $n = 4$ we can form the configurations $(t_{2g})^4$, $(t_{2g})^3e_g$, $(t_{2g})^2(e_g)^2$, etc. Each of these configurations can, in accordance with the Pauli principle, be used as a basis for a number of terms, and one could prepare a list similar to the one in Table 2.2.

According to Hund's rule, the lowest energy within each configuration would be the term with the highest multiplicity. For $(t_{2g})^4$ the highest multiplicity allowed by the Pauli principle is a triplet, whereas for $(t_{2g})^3e_g$ a quintet term is allowed. In a weak crystal field the ground state is a quintet and Table 3.2 shows that it is derived from the configuration $(t_{2g})^3e_g$.

The energy difference between corresponding terms of $(t_{2g})^4$ and $(t_{2g})^3e_g$ is the crystal field energy Δ. When Δ is less than 10,000 cm^{-1}, the higher multiplicity of the $(t_{2g})^3e_g$ terms gives them the lower energy. In strong crystal fields Δ may be as large as 30,000 cm^{-1}, and in this case the triplet term derived from the $(t_{2g})^4$ will have the lower energy. With the Pauli principle it is possible to show that this triplet has threefold orbital degeneracy, but only by group-theoretical methods can one show that it is a ${}^3T_{1g}$ term. Similar reasoning shows that low-spin complexes would be expected for strong crystal fields with up to seven $3d$ electrons. For $n = 6$ we have the interesting case of all paired electrons in strong crystal fields, and $K_4Fe(CN)_6$ is diamagnetic. Complexes containing NO, CO, and CN^- correspond to strong crystal fields, whereas those with H_2O, NH_3, Cl^-, and similar ligands usually correspond to weak crystal fields.

The crystal field orbitals t_{2g} and e_g are convenient for forming configurations for strong crystal fields, but they have some complications for weak crystal fields. Electron–electron repulsion can give a serious mixing of these configurations in weak crystal fields. In Table 3.2 the ground-state configuration of $(3d)^2$ is a mixture of $(t_{2g})^2e_g$ and $t_{2g}(e_g)^2$ in weak crystal fields. We also show that the ${}^4T_{1g}$ term in Fig. 3.5*b* is a mixture of crystal field configurations. As a result of this mixing the spacing between the ${}^4T_{1g}$ and ${}^4T_{2g}$ is not exactly Δ. We will not go into these effects in any detail.

IMPORTANCE OF ORBITAL DEGENERACY

The symbols A, E, and T give the orbital degeneracy of the terms. The A terms have no orbital degeneracy and they have many of the properties of the S terms in free atoms. The T terms have a threefold orbital degeneracy and they have many of the characteristics of P terms in free atoms. In particular, they have spin–orbit splittings and a large orbital contribution to the magnetic moment of the complex. The E terms with twofold degeneracy have no analog in the free atom, but they have no spin–orbit splittings and their magnetic properties are similar to A terms. The spin degeneracy is, of course, expressed by the multiplicity. There is a complete correspondence between the spin properties of a free atom and of a complex. This is because the crystal field is electrostatic and it only affects the orbital motion of the electrons. For A and E terms we will find that the magnetic properties of a complex are dominated by the electron spin, but in a T term the orbital motion of the electrons can still generate magnetic moments that are as large as spin moments.

DEVIATIONS FROM OCTAHEDRAL SYMMETRY

Crystal fields with cubic or tetrahedral symmetry also split the $3d$ orbitals into orbitals with t and e symmetry. The important difference with octahedral symmetry is that the e orbital is lower in energy and the t orbital is higher. As a result, the energy-level diagram in Fig. 3.5*a* is simply inverted for cubic or tetrahedral symmetry. Without to much difficulty one can also reconstruct Table 3.2 for these cases. For tetrahedral $(3d)^1$ its ground state is ^2E and for tetrahedral $(3d)^7$ it is $^4A_2$. The designation g is absent, since a tetrahedron has no center of symmetry according to the principles of group theory.

A tetragonal or trigonal field can be considered as arising by a distortion

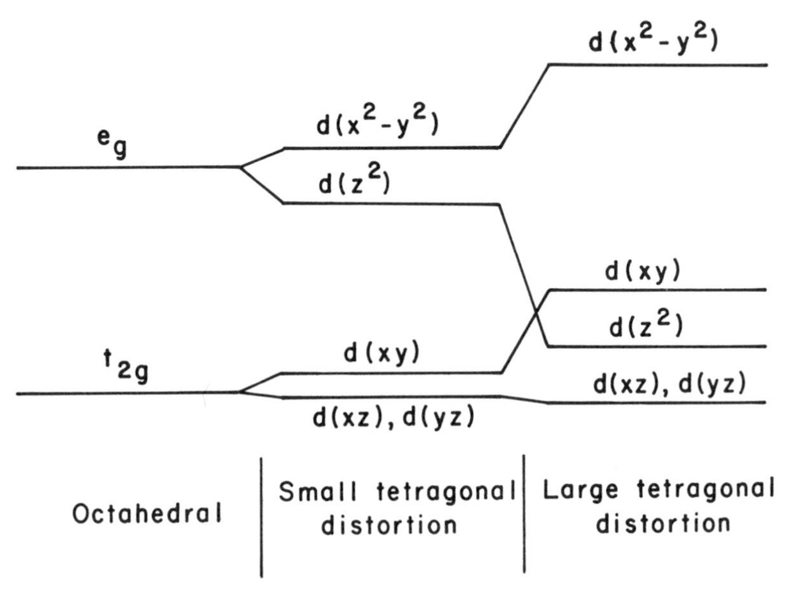

Figure 3.6. *Effect of a tetragonal distortion on the crystal field energy levels. Since this distortion stabilizes the $d(z^2)$ and destabilizes the $d(x^2 - y^2)$, it must correspond to a net decrease in the field produced by the ligands along the z axis and a smaller increase along the x and y axes.*

of an octahedral one. A tetragonal distortion along the z axis splits the t_{2g} of octahedral symmetry into a doublet, composed of the $d(xz)$ and $d(yz)$, and a single orbital, the $d(xy)$. This kind of a splitting can be understood from simple Coulombic considerations and we illustrate such splittings in Fig. 3.6. A crystal field with little or no symmetry to it will simply provide a $3d$ electron with five orbitals, all with different energy. Orbital degeneracies are only possible if the crystal field has a reasonable amount of symmetry.

SPIN-ORBIT SPLITTINGS

If the crystal field terms have both orbital and spin degeneracy one might expect some of this to be broken by spin–orbit interaction. Figure 3.5a shows that the $^2T_{2g}$ term is split by spin–orbit interaction into a lower term with fourfold degeneracy and an excited one with twofold degeneracy. The ^2E_g term in this figure is shown as unsplit by spin–orbit interaction, and it is a general result that an E term is not split. Spin–orbit interaction can give small shifts to A and E terms but no splittings.

The only spin–orbit splittings will be with the T terms. The $^2T_{2g}$ for $(3d)^1$ has a simple splitting pattern and it is the only one that we will discuss in detail. There is an analogy between its spin–orbit splitting and that of a ^2P free-ion term. For the free ion the two spin–orbit terms are $^2P_{1/2}$ and $^2P_{3/2}$. Their degeneracies are two and four. Their splitting is $3\lambda/2$. One can note the similarity with the $^2T_{2g}$ in Fig. 3.5a. The only difference is that the order must be inverted for the $^2T_{2g}$ so that the lower energy corresponds to the fourfold degenerate term.

In Section 3.1 we calculated the paramagnetic susceptibility for a free ion in a ^2P state. Although the Landé — g-value formula is not correct for the two spin–orbit terms of the $^2T_{2g}$, the final formula for its susceptibility is similar to Eq. (3.17). One finds for the $^2T_{2g}$ that μ_{eff} approaches zero for $kT \ll \lambda$ but increases to $\sqrt{5}$ for $kT \gg \lambda$. A similar variation of μ_{eff} with temperature is shown in Fig. 3.7 for $^4T_{1g}$ (Co^{+2}). The high-temperature limits are both consistent with Eq. (3.19a), and for the $^2T_{2g}$ μ_{eff} is identical to a ^2P with $S = \frac{1}{2}$ and $L = 1$. The Co^{+2} case will be discussed later.

QUENCHED ORBITAL ANGULAR MOMENTUM

The wave functions shown in Fig. 3.4 correspond to "quenched" orbital angular momentum. If electrons are in these states, they rotate against the crystal field in such a way that their angular momentum is not quantized along any axis. These functions are eigenfunctions of $\mathsf{L}^2(i)$ but not of $\mathsf{L}_z(i)$. As a result, these quenched functions will not give any orbital

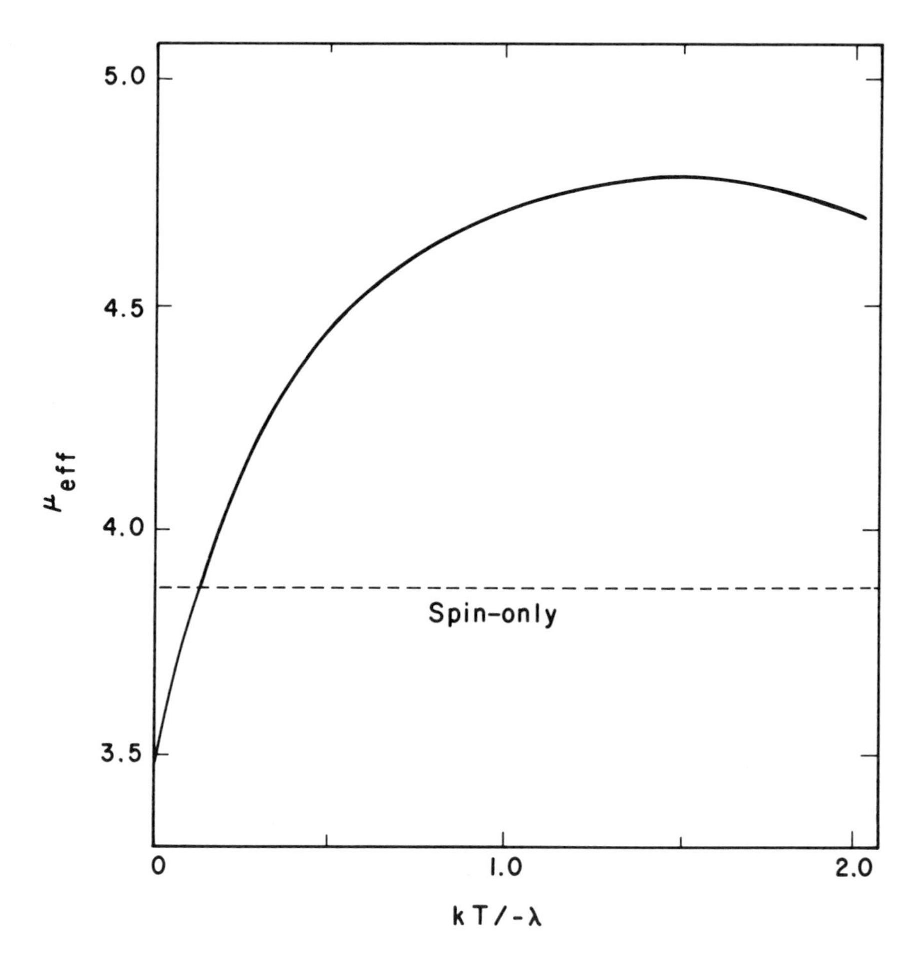

Figure 3.7. *Predicted temperature dependence of* μ_{eff} *for* Co^{+2} *in an octahedral crystal field. Griffith's calculation for a* $(t_{2g})^5(e_g)^2$ *configuration* [*Trans. Faraday Soc.*, **54,** 1109 (1958)].

contributions to $\epsilon^{(1)}$ in Eq. (1.27). As long as there is no orbital degeneracy the orbital moment can only contribute to $\epsilon^{(2)}$ and a spin-only formula is fairly accurate. If n is the number of unpaired electrons in the complex, the spin-only formula is

$$\mu_{eff} = g_0\sqrt{S(S+1)} = \sqrt{n(n+2)}. \tag{3.26}$$

The A terms have no orbital degeneracy and Eq. (3.26) is a good approximation for them. The E_g terms have a twofold orbital degeneracy, but this degeneracy does not allow an orbital contribution to $\epsilon^{(1)}$. It is common to say that an E term is nonmagnetic. This means that it has no spin–orbit splittings and no large orbital magnetic moment.

The T terms are magnetic and their orbital angular momentum is not quenched. It is possible for an electron to jump between a $d(xz)$ and a $d(yz)$ orbital and have angular momentum quantized along the z axis. It is also possible for it to jump between a $d(xz)$ and a $d(xy)$ and have angular momentum quantized along the x axis. As a result, the orbital moment can contribute to $\epsilon^{(1)}$ with such degeneracies and Eq. (3.26) has no validity for T terms.

The Cr^{+3} ion in an octahedral crystal field is a good example of quenched orbital angular momentum. As shown in Fig. 3.5b, its ground-state term is $^4A_{2g}$. It should follow a spin-only theory and agree with Eq. (3.26) with $n = 3$. One would think that the orbital moment of the $^4T_{2g}$ excited state could be completely neglected for $\Delta \sim 10{,}000\ \text{cm}^{-1}$, and kT at room temperature is only 200 cm^{-1}. As a result, the $^4T_{2g}$ term is thermally populated to only a very small degree. The $^4T_{2g}$ term will contribute to α, and in the summation of $\epsilon^{(2)}$ in Eq. (1.27) "high-frequency" terms involving the $^4T_{2g}$ will be present.

There is an additional way that the $^4T_{2g}$ excited state will contribute to the susceptibility of Cr^{+3} and it is in the evolution of $\epsilon^{(1)}$. The ground-state wave functions in Eq. (1.29) must be the correct wave functions including spin-orbit interaction. In Fig. 3.5b the ground state was labeled as $^4A_{2g}$, but this is without spin–orbit interaction. The true ground-state wave function is a mixture of $^4A_{2g}$ and $^4T_{2g}$. The fraction of $^4T_{2g}$ in the ground state is small and it is approximately equal to λ/Δ. Symmetry considerations do not allow spin–orbit interaction to mix any of the $^4T_{1g}$ into the ground state. As a result of this admixture, the magnetic moment of the ground state has an orbital contribution.

In the vector model the $^4A_{2g}$ for Cr^{+3} possesses a quenched vector **L**. This means that **L** moves in such a way that it has no average component along any axis. Its magnetic moment is present but it is not aligned in any direction. The vector **S** is aligned and it precesses around Z. If we include spin–orbit interaction, **L** is partly aligned by **S**, and the ground state for Cr^{+3} does have a small component of its orbital magnetic moment aligned along Z.

To put this contribution of the orbital moment on a quantitative basis we must use quantum mechanics. The most convenient way to do this is to use what is known as the spin Hamiltonian. This Hamiltonian is the most convenient way to include all the effects of spin–orbit interaction on the magnetic properties of ions with quenched orbital angular momentum. It was originally developed to explain EPR spectra, but it is also ideal for determining the amount of deviation from the spin-only formula for ions with A or E ground-state terms. In this chapter we will consider only the isotropic spin Hamiltonian suitable for octahedral, cubic, or tetrahedral symmetry where the energy is independent of the relative orientation of the magnetic and crystal fields. The anisotropic spin Hamiltonian will be considered in Chapter 4.

3.3. Spin Hamiltonian and Spin–Orbit Mixing

If all the terms in the atomic Hamiltonian that depend upon **L**, **S**, and a magnetic field H are gathered together, we have, neglecting spin–spin interaction,

$$\mathsf{H}(\mathbf{L}, \mathbf{S}) = \lambda \mathbf{L} \cdot \mathbf{S} + \mu_B(\mathbf{L} + 2\mathbf{S}) \cdot \mathbf{H}. \tag{3.27}$$

The effect of the crystal field on this Hamiltonian can be to remove all the first-order contributions from **L** to zero and to make **L** only contribute through second-order terms. M. H. L. Pryce (1950) showed that when this was taken into account a very general spin Hamiltonian could be formed. To second-order for an octahedral, cubic, or tetrahedral crystal field and for an A or E term or for a spin–orbit term with no orbital degeneracy,

$$\mathsf{H}(\text{spin}) = 2\mu_B(1 - \lambda\Lambda)\mathbf{S} \cdot \mathbf{H} - \lambda^2\Lambda\mathbf{S}^2 - \mu_B^2\Lambda H^2 \tag{3.28a}$$

where **S** is a new spin operator and where Λ is a constant that depends both upon the crystal field and upon the characteristics of the quenched orbital angular momentum. The allowed energy levels for Eq. (3.28a) can be readily found and

$$\epsilon(\text{spin}) = 2\mu_B(1 - \lambda\Lambda)HM_S - \lambda^2\Lambda S(S + 1) - \mu_B^2\Lambda H^2. \tag{3.28b}$$

The first term in Eq. (3.28b) is the energy of an electron spin magnetic moment with a g value that differs from 2. If the first term is written as $g\mu_B HM_S$, then $g = g_0(1 - \lambda\Lambda)$. The second term is a shift in the energy levels that is independent of M_S and H and so is an uninteresting constant. The third term is much more interesting, for it is second-order in the field and has the characteristics of an induced moment along the field direction. This can be made clear if we determine μ_Z by

$$\mu_Z = -\frac{\partial\epsilon}{\partial H} = -g\mu_B M_S + 2\mu_B^2\Lambda H \tag{3.29a}$$

so that

$$\epsilon^{(1)} = g\mu_B M_S = g_0(1 - \lambda\Lambda)\mu_B M_S \tag{3.29b}$$

$$\epsilon^{(2)} = -\mu_B^2\Lambda. \tag{3.29c}$$

If this result is used in Eq. (1.27) to determine the susceptibility of N spins and we neglect the population in excited states,

$$\chi_M = \frac{Ng^2\mu_B^2}{3kT} S(S + 1) + 2N\mu_B^2\Lambda. \tag{3.30}$$

The first term is the spin-only paramagnetism expected with quenched orbital angular momentum but with a modified g value. The second term is the second-order paramagnetism, which results from the orbital moment.

Since S_Z commutes with H(spin), the magnetic moment in Eq. (1.33) is entirely an orbital moment and with $\alpha(\text{para}) = 2\mu_B^2\Lambda$ we obtain

$$\Lambda = \frac{1}{3}\sum_{n\neq 0} \frac{|\langle\psi_0^0|\,\mathsf{L}\,|\psi_n^0\rangle|^2}{\epsilon_n^0 - \epsilon_0^0}. \tag{3.31a}$$

or

$$\Lambda = \sum_{n\neq 0} \frac{|\langle\psi_0^0|\,\mathsf{L}_Z\,|\psi_n^0\rangle|^2}{\epsilon_n^0 - \epsilon_0^0}. \tag{3.31b}$$

$Cr(H_2O)_6^{+3}$

For the ${}^4A_{2g}$ ground state of Cr^{+3} in Fig. 3.5*b*, only the ${}^4T_{2g}$ has to be used in Eq. (3.31). Exact calculations show that in this case

$$\Lambda(Cr^{+3}) = \frac{4}{\Delta}. \tag{3.32}$$

This same value can be obtained from Eq. (3.31*a*), modified in the form of Eq. (3.20), if one assumes that the quenched orbital moment of the ${}^4F$ term forms a "perpendicular" moment with $\mathbf{L}^2 = L(L + 1) = 12$.

With Eq. (3.30) and $S = \frac{3}{2}$,

$$\chi_M(Cr^{+3}) = \frac{5Ng^2\mu_B^2}{4kT} + \frac{8N\mu_B^2}{\Delta} \tag{3.33}$$

where

$$g = g_0(1 - \lambda\Delta) = 2.00 - \frac{8\lambda}{\Delta}. \tag{3.34}$$

Since Δ is much larger than kT at room temperature, the first term is the more important. With $\Delta = 17{,}400\ \text{cm}^{-1}$ and $\lambda \approx 87\ \text{cm}^{-1}$ one finds from Eqs. (3.33) and (3.34) at 293°K that

$$\chi_M(\text{Cr}^{+3} \text{ at } 293^\circ\text{K}) = (6150 \times 10^{-6}) + (120 \times 10^{-6}),$$

and $g = 1.96$.

The second term in Eq. (3.33) is seen to be small and it also comes rather close to just canceling the diamagnetism expected for Cr^{+3} and its associated anions. This fact can be seen by comparison with Table 1.2 and the second-order paramagnetism does not give any serious deviations from the Curie law. For Cr^{+3} one can simply modify Eq. (3.26) by using $g = 1.96$ in place of g_0. The theoretical μ_{eff} value for Cr^{+3} in water is $g\sqrt{15}/2$ or 3.79, which is in excellent agreement with Table 3.1.

In Table 3.3 we give the theoretical Λ values for some common ions with A and E ground-state terms. Also given in this table are approximate Δ values for the octahedral water complexes. These Δ values are derived from the optical absorption spectrum of these complex ions.

Table 3.3. *Spin-Only Ions for Weak Octahedral Crystal Fields*

Ion	Free-ion term	Octahedral term	Λ	Δ[a]
Cr^{+3}	^4F	$^4A_{2g}$	$\frac{4}{\Delta}$	17,400
Cr^{+2}	^5D	^5E_g	$\frac{2}{\Delta}$	14,000
Mn^{+2}	^6S	$^6A_{1g}$	0	8,000
Ni^{+2}	^3F	$^3A_{2g}$	$\frac{4}{\Delta}$	8,500
Cu^{+2}	^2D	^2E_g	$\frac{2}{\Delta}$	12,500

[a] Approximate values in cm^{-1} for the water complex.

$VO(H_2O)_5^{+2}$

The data given in Table 3.1 for V^{+4} are actually for the VO^{+2} ion. With five coordinated water molecules this ion would have a tetragonal crystal field with a strong axial contribution due to the O^{-2}. The lowest level in this case would be derived from the $d(xy)$ orbital and have no orbital degeneracy. This yields an orbital singlet similar to Cr^{+3} and it should have a small second-order paramagnetism and an altered g value. The EPR g value for $VO(H_2O)_5^{+2}$ is 1.962, so that the close agreement with the spin-only value in Table 3.1 is to be expected.

$Co(H_2O)_6^{+2}$

The energy-level diagram for the $(3d)^7$ configuration in an octahedral crystal field can be derived from Fig. 3.5b. This ion can be looked upon as three holes in a $(3d)^{10}$ filled shell. Since holes are positive this simply inverts the levels in Fig. 3.5b for Co^{+2}. It is also seen from Table 3.1 that holes have a negative spin–orbit interaction, so that the entire level diagram is inverted and the lowest level for Co^{+2} is a doubly degenerate one derived from the ${}^4T_{1g}$. Whereas $S = \frac{3}{2}$ for Co^{+2} without a crystal field, the lowest spin–orbit level must correspond to $S = \frac{1}{2}$ in Eq. (3.28b), for it is only doubly degenerate. The spin operator in Eq. (3.28a) is sometimes called a fictitious spin because it has the properties of a real spin angular momentum, but it often has no correspondence to the spin operator in Eq. (3.27). The g value for the lowest state of Co^{+2} is quite a bit larger than 2.0 because λ is negative and Λ involves the close-by spin–orbit levels. This also gives a large $\epsilon^{(2)}$ value. Since $\lambda \approx kT$ at room temperature, the other spin–orbit levels must be included in Eq. (1.27), and one finds that μ_{eff} will vary with temperature. A plot of a predicted variation of μ_{eff} for Co^{+2} is shown in Fig. 3.7.

It can be seen in this figure that when $|\lambda| = kT$, μ_{eff} approaches the measured solution value of 5.04. The calculation in Fig. 3.7 assumes that the configuration of Co^{+2} is $(t_{2g})^5(e_g)^2$. It can be seen in Table 3.2 that this is only approximately correct in a weak crystal field because of the mixing of the ${}^4T_{1g}$ levels. If this mixing is taken into account, then μ_{eff} is increased, so that the 5.04 value can be explained. When $kT \gg |\lambda|$, Eq. (3.19a) predicts that $\mu_{\text{eff}} \rightarrow \sqrt{17} = 4.12$, and Fig. 3.7 does approach this value for high temperatures. It is very clear that for Co^{+2} the Curie law will not be followed and that even the Curie–Weiss law must only be a rough approximation for the variation of χ_M with temperature. It is also clear from Fig. 3.7 that the spin-only value given by Eq. (3.26) has no validity for Co^{+2} in a weak octahedral crystal field.

From Table 3.2 it can be seen that $V(H_2O)_6^{+3}$ and $Fe(H_2O)_6^{+2}$ having ground-state T terms must be similar to $Co(H_2O)_6^{+2}$, but we will not discuss these ions in any detail.

LANTHANIDE IONS

These ions have small crystal field interactions. The smallness of these interactions is due to the fact that the $4f$ orbitals have a rapidly decreasing radial wave function and are shielded by the filled d shells from the crystal fields. It can be seen from Table 3.1 that except for Sm^{+3} and Eu^{+3} the crystal fields seem to have little effect upon μ_{eff}. For Sm^{+3} and Eu^{+3} Van Vleck pointed out that the spin–orbit spacing† is close to kT, so that the ${}^6H_{3/2}$ and ${}^7F_1$, respectively, must also be populated and contribute to μ_{eff}. Detailed calculations show that this is the case for these ions. The first few lanthanide ions do not seem to follow the Curie law, presumably because of contributions from other terms or to a second-order paramagnetism. The use of the Curie–Weiss law is not particularly justified for ions in solution, but it is often used to explain any variation in μ_{eff} not understood by the investigator. In the case of a diamagnetic ground state and a low-lying paramagnetic excited state, Δ in the Curie–Weiss law can be related to the energy difference between these states, but in most cases Δ is only an additional empirical constant introduced to better fit any variation of μ_{eff} with temperature.

SUMMARY

We have seen that the magnetic properties of the paramagnetic metal ions in solutions and in complexes is a very complex subject indeed. We have only been able to scratch the surface of this topic, but we do hope that you have been able to see the use of Eq. (1.27) for these compounds. Electron paramagnetic resonance is a very powerful tool, and since it only depends upon the first term of Eq. (3.28) it has some advantage over the measurement of bulk susceptibility.

References

C. J. Ballhausen, *Introduction to Ligand Field Theory*, McGraw-Hill Book Company, New York (1962). An excellent book, which includes some magnetic properties.

† These ions will only approximately follow Eq. (2.20) because of deviations from LS coupling.

T. M. Dunn, D. S. McClure, and R. G. Pearson, *Some Aspects of Crystal Field Theory*, Harper and Row, Publishers, New York (1965). Chaps. 1 and 3 by Dunn are a good supplement to this material.

B. N. Figgis, *Introduction to Ligand Fields*, John Wiley & Sons, Inc. (Interscience Division), New York (1966). More emphasis on magnetic properties than either Ballhausen or Orgel.

J. S. Griffith, *The Theory of Transition-Metal Ions*, Cambridge University Press, New York (1961). For those who have mastered Condon and Shortley.

E. König, *Magnetic Properties of Coordination and Organo-Metallic Transition Metal Compounds*, Landolt–Börnstein New Series Group II, Vol. 2, Springer-Verlag New York, Inc., New York (1966). A fairly recent compilation for transition metals including EPR results.

L. E. Orgel, *An Introduction to Transition–Metal Chemistry: Ligand–Field Theory*, Methuen and Co. Ltd., London (1960). A good short introduction.

Problems

3.1. Show that Eq. (3.5) gives the answer to Problem 1.4.

3.2. Make a table showing the values of μ_{eff} for the spin-only case with 1, 2, 3, 4, and 5 unpaired electrons.

3.3.* Use the data in Table 3.1 to predict a g value for $Cu(H_2O)_6^{+2}$. The EPR value is 2.20. Also calculate a theoretical value from the data given in Table 3.3.

3.4. Draw a curve showing how μ_{eff} for the two ^2P states as predicted by Eq. (3.17) would vary with $1/x$. This curve should have some similarity to Fig. 3.7.

3.5. Modify Eq. (3.17) for the case that λ is negative so that the $^2P_{3/2}$ term is lower in energy than is the $^2P_{1/2}$. What value of μ_{eff} would result in this case at high temperatures? Is this consistent with Eq. (3.19*a*)?

3.6. Predict which iron-group ions would have large changes in susceptibilities when their octahedral crystal fields change from weak to strong. Which ions should have essentially no change in susceptibilities under these circumstances?

3.7.* Predict the lowest energy configurations for the iron-group ions in weak and strong tetrahedral crystal fields.

3.8. The reported susceptibility of $K_4Mn(CN)_6 \cdot 3H_2O$ at 291°K is $10^6\chi_M = +1906$. What μ_{eff} does this give? What configuration does this establish for the Mn^{+2}? Explain why μ_{eff} is not exactly the spin-only value and predict its high-temperature limit.

3.9. The reported susceptibility of $K_2CaCo(NO_2)_6$ at 294°K is $10^6\chi_M = +1414$ and if one corrects for the diamagnetic part of α, then $10^6\chi_M(\text{para}) = +1541$. Determine μ_{eff} both with and without the diamagnetic correction. What is the configuration of Co^{+2} in this case? Without determining α(para), which μ_{eff} value agrees better with the predicted spin-only value?

Solutions:

3.3. With the equation $\mu_{\text{eff}} = g\sqrt{S(S+1)}$ and the data in Table 3.3 we obtain

$$g = \frac{1.94 \times 2}{\sqrt{3}} = 2.24.$$

With the equation $g = g_0(1 - \lambda\Lambda)$ and the data in Tables 3.1 and 3.3 we obtain

$$g = 2.00\left(1 + \frac{852 \times 2}{12{,}500}\right)$$

$$= 2.00(1 + 0.136) = 2.27.$$

3.7. If we neglect any mixing of the weak-field configurations by electron–electron repulsion, we can construct the following table for tetrahedral crystal fields:

Electrons	Weak	Strong
1	e	e
2	$(e)^2$	$(e)^2$
3	$(e)^2t_2$	$(e)^3$
4	$(e)^2(t_2)^2$	$(e)^4$
5	$(e)^2(t_2)^3$	$(e)^4t_2$
6	$(e)^3(t_2)^3$	$(e)^4(t_2)^2$
7	$(e)^4(t_2)^3$	$(e)^4(t_2)^3$
8	$(e)^4(t_2)^4$	$(e)^4(t_2)^4$
9	$(e)^4(t_2)^5$	$(e)^4(t_2)^5$

4

Molecular Paramagnetism

IN CHAPTERS 1 through 3 we saw that the basic spherical symmetry of atoms leads to a fairly regular energy-level scheme such that many of the orbitals in atoms have a high degeneracy. This degeneracy together with the general accuracy of Hund's rule often causes the ground states of atoms to have unpaired spins and hence be paramagnetic. In crystal field theory we have seen that weak crystal fields nearly destroy the orbital contributions to paramagnetism in atoms but that they can still retain the basic high spin. In most molecules, however, we must consider the atoms as being subjected to strong crystal fields, for most molecules are truly covalently bonded. In addition, the symmetry of most molecules is rather low. The combination of low symmetry and strong crystal fields leads to molecules that most often have essentially neither the orbital nor the spin paramagnetism of the original atoms. For molecule with an even number of electrons these factors lead most commonly to diamagnetic ground states.

4.1. Loss of Orbital Degeneracy

In the oxygen atom the ground state has the configuration $(1s)^2(2s)^2(2p)^4$. The $2p$ orbitals are unfilled, for with $l = 1$ there is an orbital degeneracy of three and they could hold six electrons all together. The application of the Pauli principle to $(np)^4$ gives ^1S, ^1D, and ^3P terms with Hund's rule predicting, correctly, that the ^3P is the ground state. In the free O atom both orbital and spin moments would contribute to its paramagnetism.

If the O atom is part of a linear molecule, its energy levels can be deduced from a crystal field model. The CH_2 molecule with H—C—H geometry can be approximated as an O atom with the two protons providing an axial crystal field. Its $2p$ orbitals under the influence of such a field is split into a σ orbital, which would point toward the protons, and π orbitals, which point away from them. There is no orbital degeneracy to a σ orbital, but there are two degenerate π orbitals corresponding to clockwise and counterclockwise rotation about the H—C—H axis. The angular momentum of these π electrons is quantized along this molecular z axis as $\pm\hbar$. Since the protons are positive, the σ(or $2p_z$) orbital has the lower energy.

The one-electron energy levels for an O atom and a linear CH_2 molecule are given in Fig. 4.1. One can see that the $2p\pi$ orbitals would be half filled with the last two electrons. If these two electrons were in the same π orbital, they would rotate together around the H—C—H axis, but the Pauli principle would require them to have paired M_S values. They would form a $^1\Delta$ term, where Δ means an angular momentum of $\pm 2\hbar$ quantized along the H—C—H axis. If the two electrons were in different π orbitals, they would form either $^1\Sigma$ or $^3\Sigma$ terms. There would be no net angular momentum along the H—C—H axis, but the Pauli principle would now allow either paired or unpaired spins.

Since the $^1\Delta$, $^1\Sigma$, and $^3\Sigma$ terms arise from the same configuration, we can use Hund's rule to predict the ground state. It suggests that the $^3\Sigma$ should be the lowest-energy term, as indicated in Fig. 4.1. One can see that in a linear CH_2 the orbital moment of the O atom is quenched but the spin moment remains. In a bent CH_2 molecule one should think in terms of molecular orbitals, but the most important fact is that the degeneracy of the two π orbitals is broken. Figure 4.1 indicates that one of these π oribtals may be even lower in energy than the σ. The new orbital symbols a and b refer to their symmetries with respect to 180° rotation about the CH_2 axis. One orbital is primarily a $2p_x$ and the other a $2p_y$.

A bent CH_2 is shown in Fig. 4.1 as having its last two electrons in the

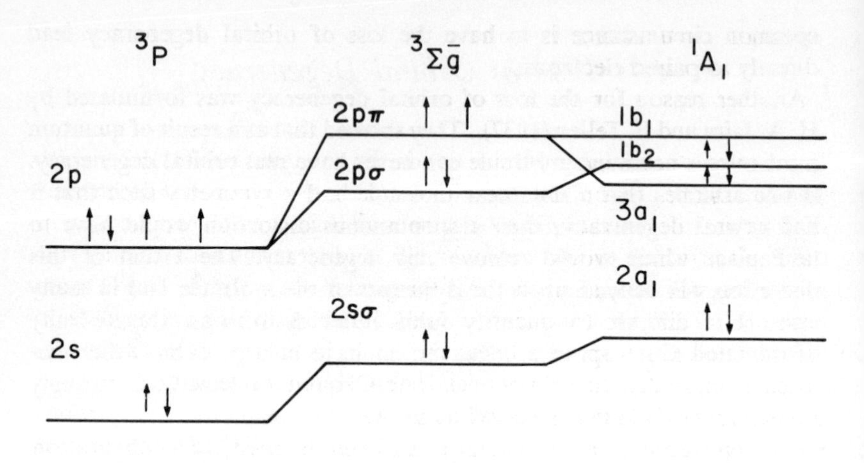

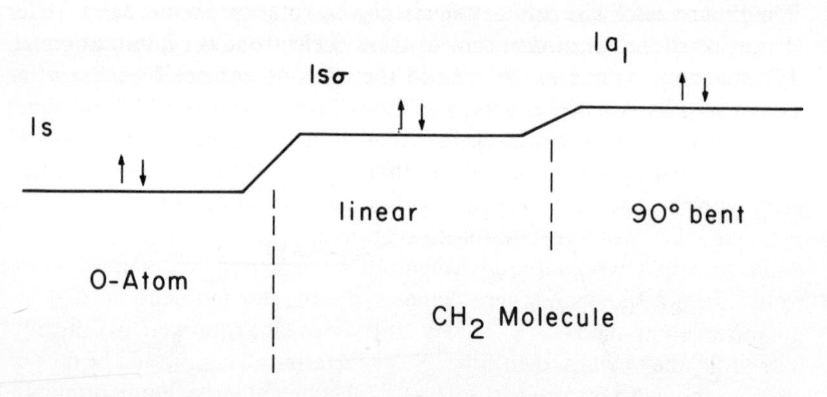

Figure 4.1. *Ground-state electronic configuration for isoelectronic O-atom and* CH_2 *molecule. (After G. Herzberg and Walsh). The arrows show the relative Ms values of the electrons.*

$1b_2$. They would have to be spin paired and form a singlet term. A great deal of chemical evidence indicates that CH_2 can be formed, and can do chemical reactions, as either a triplet or singlet species. For a number of years it was supposed that triplet CH_2 was linear and singlet CH_2 was bent. These same experiments indicated that the triplet CH_2 had the lower energy. More recently, theoretical calculations indicate that the lowest energy for CH_2 corresponds to a slightly bent species with unpaired electrons ($^3B_1$). This means that CH_2 is equivalent to a high-spin complex with only a small energy difference between half-filled one-electron orbitals. It is an unusual situation for molecules, and the more common circumstance is to have the loss of orbital degeneracy lead directly to paired electrons.

Another reason for the loss of orbital degeneracy was formulated by H. A. Jahn and E. Teller (1937). They showed that as a result of quantum mechanics a nonlinear molecule can never have real orbital degeneracy. If one assumes that a nonlinear molecule had a symmetry such that it had orbital degeneracy, then a spontaneous distortion would have to take place which would remove this degeneracy. The extent of this distortion will depend upon the dynamics of the molecule and in many cases it is difficult to quantify Jahn–Teller distortions. Octahedrally coordinated Cu^{+2} appears, however, to have a large Jahn–Teller distortion, and such complexes usually are found to have four strongly bound ligands and two weakly bound ones.

The application of a octahedral crystal field to the $(3d)^9$ configuration of Cu^{+2} can be obtained from Fig. 3.5*a*. One hole in a filled $(3d)^{10}$ configuration inverts the $(3d)^1$ energy-level diagram so that the ground state is e_g or a ^2E_g term. The excited state would be $^2T_{2g}$, as shown in Fig. 4.2. The ground state has orbital degeneracy and a spontaneous Jahn–Teller distortion of the octahedral coordination occurs to lower it to tetragonal. In tetragonal symmetry the ground state has no orbital degeneracy, as shown in Fig. 4.2. Of course, the Jahn–Teller distortion has no direct

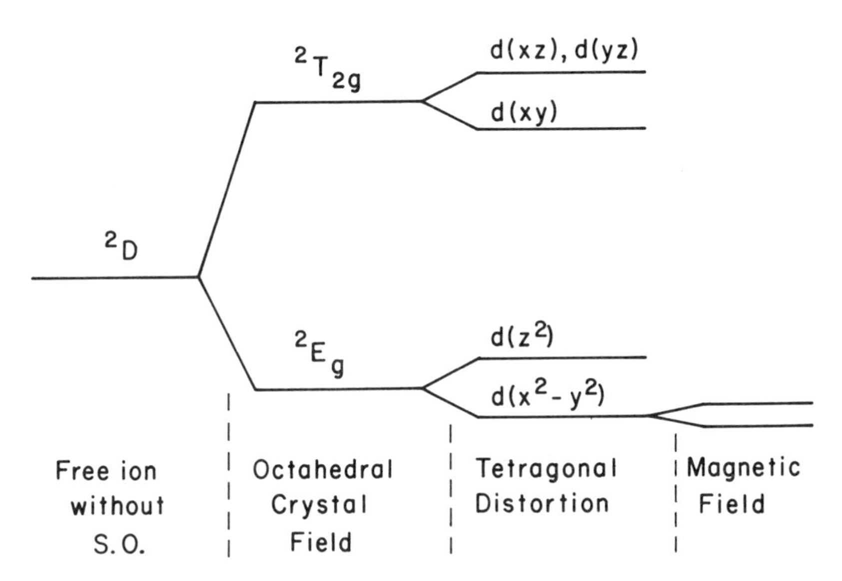

Figure 4.2. *Energy levels for* $Cu^{+2}(3d)^9$. *The observed complexes have an elongated tetragonal structure. This structure stabilizes the positive holes in the* $d(x^2 - y^2)$ *orbital and removes the orbital degeneracy in the ground state. This is the hole analogue of Fig.* 3.6.

effect upon the unpaired spin and so Cu^{+2} is still paramagnetic. The g values of the ground state are affected by this distortion and the axial ligands become more weakly bound.

An important consequence of the Jahn–Teller theorem is that the spin-Hamiltonian formalism can be applied to almost all paramagnetic systems. A restriction on the use of Eq. (3.28) was that the ground state have no orbital degeneracy. In almost all systems one finds that orbital degeneracy is sufficiently removed so that an appropriate spin Hamiltonian can be formulated to explain the magnetic splitting of energy levels. It is even found that when linear molecules are placed in solids or liquids, their orbital angular momentum is quenched by surrounding molecules; as a result, a spin Hamiltonian also explains their paramagnetism quite well.

4.2. Odd Molecules—Radicals

Another important theorem was formulated by H. A. Kramers (1930). He showed that the electrostatic forces that exist in molecules can never remove the twofold spin degeneracy in a system containing an odd number of electrons. Thus if a molecule has an odd number of electrons it must always be paramagnetic. Odd molecules must have electronic spin degeneracy with $S = \frac{1}{2}, \frac{3}{2}$, etc. The most common situation is $S = \frac{1}{2}$ forming a doublet state. External magnetic fields, of course, split such doublet levels, but according to Kramer's theorem the internal electric fields in a molecule cannot do this.

It is an unfortunate fact of chemistry that most molecules with an odd number of electrons are chemically unstable, and H atom, OH, CN, CH_3, etc., react readily with both themselves and other molecules. They are often called radicals, for they are most commonly found not as free molecules but as parts of more stable ones. Thus Kramer's theorem makes odd molecules paramagnetic, but chemistry says that they are radicals and should be unstable. A number of stable radicals, such as NO, NO_2, and ClO_2, do exist, but it is true that most odd molecules are unstable. It is the reactivity of these molecules, however, that makes them important and much work has been done to study them. The paramagnetism of radicals is an important tool in such studies and EPR is a sensitive detector for radicals.

Since the majority of molecules are covalently bonded, Eqs. (3.28) and (3.31) predict g values close to 2.0. This is because their excited states have a high energy and Λ is small. For most molecules, however, we must modify these equations because they are derived with the assumption of octahedral symmetry. For a molecule with only a twofold axis of symmetry or even lower symmetry there are three possible g values. For the general case one must have g_x, g_y, and g_z values. If we modify Eq. (3.28) for this fact and write down only those terms that depend upon both **S** and **H**, then

$$\mathsf{H}(\text{spin}) = \mu_B(g_x \mathsf{S}_x H_x + g_y \mathsf{S}_y H_y + g_z \mathsf{S}_z H_z). \tag{4.1}$$

For the covalent bonding that is characteristic of most molecules, this spin Hamiltonian should be derived starting with one-electron wave functions and spin–orbit interaction in the form of Eq. (2.22). The theoretical relations for the g values will not be given in this form, for they require a detailed knowledge of both the occupied and unoccupied one-electron orbitals and few approximations can be made. In Eq. (4.1), x, y, and z are fixed in the molecule and are defined as the principal axis system, so that g_{xy} and all similar terms are zero. For a molecule with elements of symmetry the principal axes must lie along these elements of symmetry. In the NO_2 radical, for example, there is a twofold axis of symmetry and two planes of symmetry (point group C_{2v}). The principal axes are completely defined by these symmetry elements and one can take z as along the twofold axis, y in the molecular plane, and x at right angles to this plane. The exact identification of x, y, and z is, of course, arbitrary, but the general location is determined by symmetry.

For NO_3 and similar molecules with a threefold axis of symmetry, it is possible to show that $g_x = g_y$ if z is defined as the threefold axis. This kind of symmetry forms the basis of an axial spin Hamiltonian, and for this special case Eq. (4.1) can be simplified with $g_z = g_{\parallel}$ and $g_x = g_y = g_{\perp}$:

$$\mathsf{H}(\text{axial spin}) = \mu_B g_{\parallel} \mathsf{S}_z H_z + \mu_B g_{\perp}(\mathsf{S}_x H_x + \mathsf{S}_y H_y). \tag{4.2}$$

To illustrate the kind of g values that have been observed for some simple radicals a representative list is given in Table 4.1. These values were all determined by EPR spectroscopy, for only by such methods is it practical to obtain this much detailed information on unstable radicals. Organic radicals, as we shall see, can also be studied by EPR spectroscopy, but in many cases the nuclear hyperfine terms that we have neglected in Eqs. (4.1) and (4.2) dominate the observed spectrum.

The energy levels that result from Eq. (4.1) can be easily solved when H is along either the x, y, or z axes. In these cases only one term remains, and if S_Z is taken as quantized, as usual, then

$$\epsilon^{(1)} = g_i \mu_B M_S \tag{4.3}$$

Table 4.1. *Principal g Values for Some Inorganic Radicals*[a]

Molecule	g_x	g_y	g_z
O_2^-[b]	2.00	2.00	2.175
ClO[b]	1.9909	1.9909	2.0098
CO_2^-	2.0032	1.9975	2.0014
NO_2	2.0057	1.991	2.0015
ClO_2	2.0036	2.0183	2.0088
NO_3	2.025	2.025	2.005
ClO_3	2.008	2.008	2.007

[a] Taken from Chap. 7 in *Free Radicals in Inorganic Chemistry*, by M. C. R. Symons, Advances in Chemistry Series 36, American Chemical Society, Washington, D.C. (1962). The z axis is the axis of highest symmetry and the y axis lies in the plane of the molecule.

[b] These species have $^2\Pi$ ground states and should have interesting gas-phase EPR spectra. In solids or in solution the neighboring molecules quench the orbital moment and the g values are only a little out of the ordinary. These g values must therefore depend a great deal on the interactions with the lattice.

where g_i is g_x, g_y, or g_z. If we neglect both the second-order paramagnetic and diamagnetic terms, then for $S = \frac{1}{2}$

$$\chi_M^i = N \frac{g_i^2 \mu_B^2}{4kT}. \tag{4.4}$$

Thus the susceptibility will be anisotropic. The average value of the susceptibility in all three directions is what would be measured for a powder or a liquid, and so

$$\chi_M^{\text{av}} = N \frac{g^2 \mu_B^2}{4kT} \tag{4.5a}$$

where

$$g^2 = \frac{g_x^2 + g_y^2 + g_z^2}{3}. \tag{4.5b}$$

In the axial case if H is taken so that it is at an angle θ with respect to z, one can use Eqs. (4.3) and (4.4) if

$$g_i^2 = g_\parallel^2 \cos^2\theta + g_\perp^2 \sin^2\theta. \tag{4.6}$$

Similar but more complicated equations can be written for H not along x, y, or z in Eq. (4.1).

For single crystals it is unlikely that orientations of H can be found so that it is aligned along the principal axes of every paramagnetic center. Most solid lattices contain several sets of paramagnetic centers and each set, most probably, has principal axes that are at angles with respect to each other set. The susceptibilities along the principal axes of the crystal's susceptibility tensor can be designated $\overset{1}{\chi}_M$, $\overset{2}{\chi}_M$, and $\overset{3}{\chi}_M$. For cubic, tetragonal, orthorhombic, and hexagonal crystals these principal axes must correspond to the crystal a, b, and c axes. For rhombohedral and triclinic crystals the principal axes of the magnetic susceptibility tensor may not at all correspond to the crystal axes. However, if the relative orientation of the principal axes of each paramagnetic center is known in the crystal a, b, and c axes, then $\overset{1}{\chi}_M$, $\overset{2}{\chi}_M$, and $\overset{3}{\chi}_M$ can be calculated. In only a few cases do we know enough about the detailed orientation of the x, y, and z axes in the a, b, and c system to make these calculations. Problem 4.5 illustrates this type of calculation.

4.3. Triplet States—Biradicals

As we said earlier most molecules with an even number of electrons have $S = 0$ and diamagnetic ground states. In a few cases, however, molecular symmetry and other factors can lead to two unpaired electrons and $S = 1$. The most common example of a ground-state triplet is the O_2 molecule. It is also quite probable that the first excited states of many molecules are triplets; the $(1s)(2s)$ state of He atom is an example of this.

Without spin–orbit interaction, spectroscopic transitions are forbidden from triplet to singlet states. Under these circumstances, the lowest triplet state in many molecules are metastable. The low spin–orbit interaction in the light elements makes their triplet states long-lived, and their lifetimes for radiation to the ground singlet state can be several seconds.

The technique for producing appreciable concentrations of excited triplet states is illustrated in Fig. 4.3. An intense light source close to the normal singlet–singlet absorption produces molecules in excited singlet states. By radiationless transitions some fraction of these molecules rapidly changes to the triplet system and then falls to the lowest triplet state. Once in this lowest triplet state these molecules are metastable and slowly decay to the ground singlet. The crossover from the singlet to the triplet system must take place via close coincidence of the excited vibrational states of both systems. The details of this intersystem transfer is a difficult but important subject.

With aromatic hydrocarbons like benzene and naphthalene the triplet lifetimes in frozen glasses can be close to 1 second. With good light

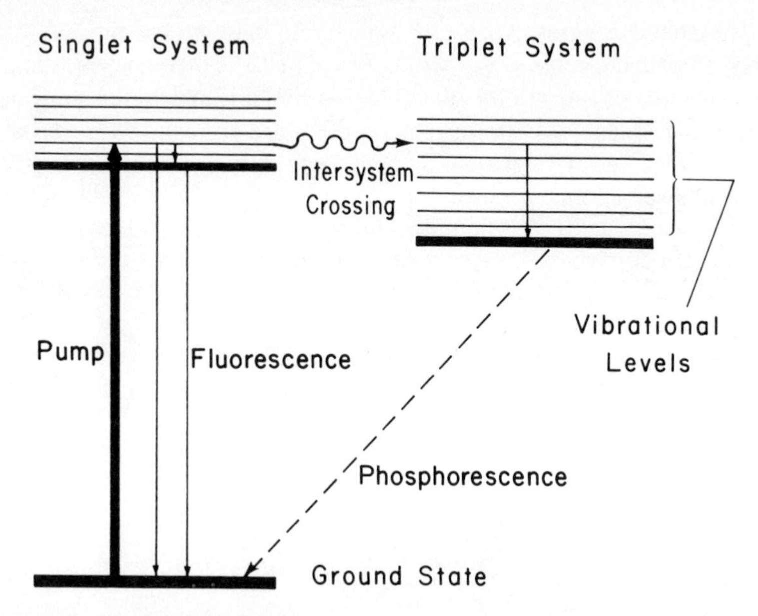

Figure 4.3. *Photochemical production of metastable triplet states. Intense light of the normal singlet–singlet absorption produces a small fraction of triplets owing to intersystem transfer.*

sources about 10 per cent of the aromatic molecules can be populated, in steady state, as triplets. It is also possible by chemistry to prepare unusual chemical species, like CH_2, which are triplets in the ground state. In some larger organic molecules there seems to be a localization of each unpaired electron in a separate place in the molecule. These species are called biradicals, but, of course, they have $S = 1$ and are only a special class of triplets.

The spin Hamiltonian for triplets must contain the interaction between the unpaired electrons. This interaction has two sources: (1) direct dipole–dipole interaction between the electrons and (2) indirect interaction via spin–orbit interaction and the quenched orbital angular momentum. For molecules made up of the first 10 elements the direct dipole–dipole interaction is important, but the heavier elements can give quite a large interaction between the unpaired electrons because of spin–orbit interaction. The common form of the spin Hamiltonian for triplets is

$$\mathsf{H}(\text{spin}) = \mu_B(g_x \mathsf{S}_x H_x + g_y \mathsf{S}_y H_y + g_z \mathsf{S}_z H_z) + D[\mathsf{S}_z^2 - \tfrac{1}{3}S(S+1)] + E(\mathsf{S}_x^2 - \mathsf{S}_y^2) \tag{4.7}$$

where D and E are constants and $S(S+1) = 2$ for triplets.

The D and E terms express the interaction between the two electrons. For a linear molecule or one with a threefold or higher axis of symmetry we have the characteristics of an axial spin Hamiltonian,

$$\mathsf{H}(\text{axial spin}) = \mu_B g_{\parallel} \mathsf{S}_z H_z + \mu_B g_{\perp}(\mathsf{S}_x H_x + \mathsf{S}_y H_y) + D[\mathsf{S}_z^2 - \tfrac{1}{3}S(S+1)]. \tag{4.8}$$

In this case the E term is zero by symmetry. Since the D term provides a splitting in the absence of a magnetic field it is often called the zero-field splitting. In true biradicals D should be quite small but in some triplets the zero-field splitting is the dominant term for normal magnetic fields. Table 4.2 gives some values for D and E in some triplet states.

Table 4.2. *Zero-Field Splitting Constants for Some Triplet States*

Molecule	D/hc (cm^{-1})	E/hc (cm^{-1})
Ground states		
O_2	+3.968[a]	0
SO	+10.552	0
S_2	~23.4	0
NH	+1.86	0
NCN	+1.568	0
CH_2	±0.688	0.0035
$C(C_6H_5)_2$	±0.405	0.019
HCC_6H_5	±0.518	0.024
Excited states		
Benzene	~0.18	~0[b]
Naphthalene	+0.100	0.015
Anthracene	±0.072	0.007
Phenanthrene	+0.100	0.047
Triphenylene	±0.134	0
Pyrene	±0.0658	0.0316
Quinoline	±0.103	0.016
Isoquinoline	±0.100	0.012

[a] For diatomic gases one often finds a constant called λ tabulated, where $D = 2\lambda$.

[b] Because of Jahn–Teller interaction triplet benzene may not be a regular hexagon.

The energy levels predicted by Eq. (4.8) can only be easily solved when H is along z. In this case the zero-field term is $D/3$ for $M_S = \pm 1$ and $-2D/3$ for $M_S = 0$. When H is not along z, Eq. (4.8) gives a quadratic energy expression. Equation (4.7) gives a quadratic even when H is along z. Figure 4.4 shows an energy-level diagram for a typical organic triplet state. It can be seen for $D = 0.100\ \text{cm}^{-1}$ that fields of over 1000

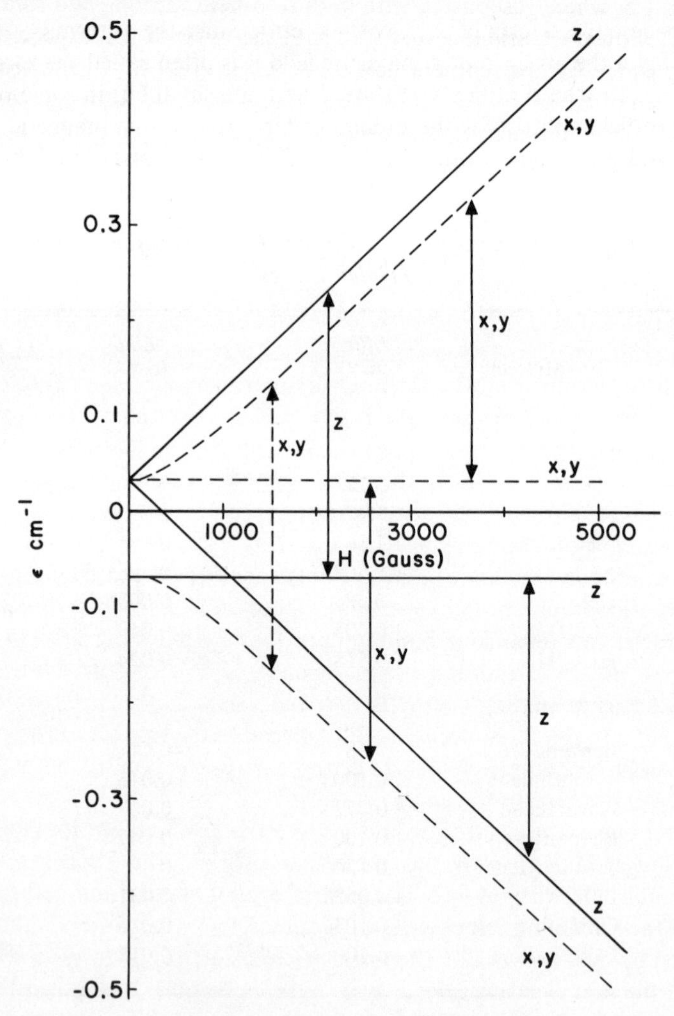

Figure 4.4. *Triplet-state energy levels with $D = 0.1\ cm^{-1}$ and $E = 0$. Solid lines are for H along z and the dashed ones are for H along either x or y. A g value equal to the free electron is assumed. The arrows show possible EPR transitions, which are discussed in Chapter* 8.

gauss are required to establish the dominance of the Zeeman effect over the zero-field splitting.

At temperatures where $kT \approx D$, triplet-state molecules should not follow the Curie law, but when $kT \gg D$ they should. This reasoning follows from Van Vleck's arguments involving the two terms in Eq. (1.27). At high temperatures the zero-field splitting becomes a "low-frequency" term and the Curie law is followed. This is clearly illustrated in Fig. 1.5, where gaseous O_2 with both zero-field splitting and rotational levels follow the Curie law very well with the value of μ_{eff} expected for a triplet state. At low temperatures when kT is close to D, the susceptibility can often be roughly fit by the Curie–Weiss law. Low-temperature susceptibility measurements are one of the best ways of establishing the sign of D, but this has been done in only very few cases.

4.4. Molecules O_2 and NO

These two molecules are excellent examples of stable diatomic molecules with paramagnetism. An understanding of their paramagnetism exposes our whole picture of molecular structure in these fundamental molecules.

In Fig. 4.1 it is seen how the $1s$, $2s$, and $2p$ orbitals of an atom are affected by a strong axial electric field. In N_2, NO, O_2, and similar molecules one can consider their electrons in molecular orbitals which are formed from combinations of the perturbed $1s$, $2s$, and $2p$ orbitals of each atom. Both bonding and antibonding molecular orbitals can be formed so that two $1s\sigma$ atomic orbitals can form a $1s\sigma$ (bonding) and $1s\sigma^*$ (antibonding) molecular orbital. Thus the $1s$, $2s$, and $2p$ orbitals on two nuclei can form four bonding and four antibonding orbitals from the $1s\sigma$, $2s\sigma$, $2p\sigma$, and $2p\pi$ atomic orbitals. For molecules close to N_2 the order of increasing energy is $1s\sigma$, $1s\sigma^*$, $2s\sigma$, $2s\sigma^*$, $2p\pi$, $2p\sigma$, $2p\pi^*$, and $2p\sigma^*$. In the N_2 molecule with 14 electrons the first six of these orbitals would be completely filled, giving a strongly bonded diamagnetic molecule.

In NO one extra electron would go into the $2p\pi^*$ and in O_2 two extra electrons would go into this molecular orbital. The electrons in the $2p\pi^*$ molecular orbital have one unit of orbital angular momentum and are either rotating clockwise or counterclockwise around the axis formed by the two nuclei. The one extra electron in NO forms a $^2\Pi$ term where the odd electron gives a spin multiplicity of 2, and it is a Π state with one unit of orbital angular momentum around the internuclear axis. It is not a P state because the axial electric field of the molecule quenches all the orbital angular momentum other than that corresponding to rotation around the molecular axis.

In the O_2 molecule the two electrons can have opposed angular momentum and form either $^3\Sigma$ or $^1\Sigma$ terms or add their orbital angular momenta and form only a $^1\Delta$ term. The Pauli principle requires that if their angular momenta do add, their spins must be opposed. The application of Hund's rule to molecules would state that the $^3\Sigma$ would have the lowest energy since it has the highest multiplicity. It is correct in this case and the observed energy-level pattern for O_2 is shown in Fig. 4.5. It can be

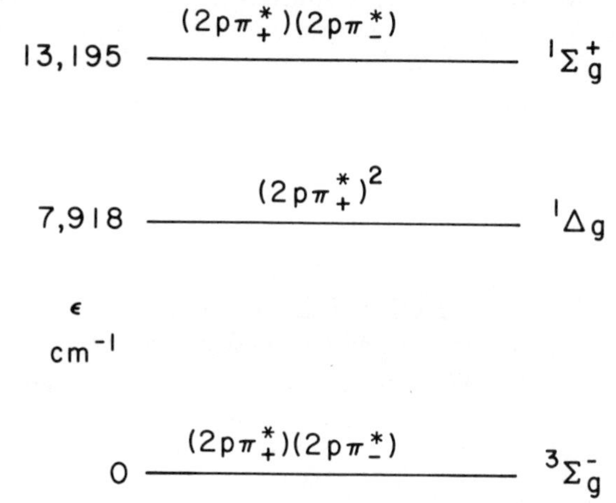

13,195 — $(2p\pi_+^*)(2p\pi_-^*)$ — $^1\Sigma_g^+$

7,918 — $(2p\pi_+^*)^2$ — $^1\Delta_g$

ϵ

cm^{-1}

0 — $(2p\pi_+^*)(2p\pi_-^*)$ — $^3\Sigma_g^-$

Figure 4.5. *Three lowest electronic energy levels for* O_2 *which arise from two electrons in the* $2p\pi^*$ *orbitals.*

seen that the ground electronic state of O_2 is a triplet state. The $^1\Delta$ state is particularly interesting, for it is a metastable excited singlet state, and in ordinary glass apparatus it has several seconds of lifetime in the gas phase at low pressure. The $^3\Sigma$ state has essentially fully quenched orbital paramagnetism and so the simple paramagnetism of a triplet state should be observed for O_2 in its ground electronic state.

Since the $^2\Pi$ term for the ground state of NO has both spin and orbital contributions to paramagnetism, it is relatively complicated. The understanding of its paramagnetism is, in fact, a high point of Van Vleck's general theories. Spin–orbit interaction in diatomic molecules is, however, relatively simple. The internuclear axis acts as a z, or molecular fixed, axis of quantization. The Hamiltonian for spin–orbit interaction has only one important term,

$$\begin{aligned} \mathsf{H}(\text{spin–orbit}) &= \lambda \mathbf{L} \cdot \mathbf{S} \\ &\approx \lambda \mathsf{L}_z \mathsf{S}_z. \end{aligned} \tag{4.9}$$

Since L_z and S_z can be taken as quantized, we can define new quantum numbers Σ and Λ such that

$$\mathsf{L}_z\psi = \Lambda\psi \qquad \text{and} \qquad \mathsf{S}_z\psi = \Sigma\psi, \tag{4.10a}$$

so that

$$\epsilon(\text{spin–orbit}) = \lambda\Lambda\Sigma. \tag{4.10b}$$

For a $^2\Pi$ term one has $\Lambda = \pm 1$ corresponding to clockwise and counter-clockwise rotation of one π electron, and since $S = \frac{1}{2}$, then $\Sigma = \pm\frac{1}{2}$. One should not be too confused by the use of Σ both as a term symbol corresponding to $\Lambda = 0$ and as the eigenvalue of S_z.

As a result of spin–orbit interaction the $^2\Pi$ term is split into two terms, called $^2\Pi_{3/2}$ and $^2\Pi_{1/2}$, which differ by λ in energy. Figure 4.6*a* illustrates the vector-model diagrams for this situation. The vector **L** is not shown since under these circumstances all the components of **L** other than $L_z = \Lambda$ are assumed as fully quenched. When the NO molecule starts to rotate, the space-fixed Z axis also becomes important. If it is more accurate to consider **S** as quantized about z, this is called Hund's case A, as illustrated in Fig. 4.6*b*. But for high-rotational angular velocities **S** will uncouple from the z axis and Fig. 4.6*c* illustrates Hund's case B. The NO molecule follows case A rather well. The spin–orbit constant $\lambda = 124\ \text{cm}^{-1}$ and the $^2\Pi_{1/2}$ term is lower in energy by essentially this value. Hund's two cases are important, for they determine the relative orientations of the orbital and spin magnetic moments. The combination of rotational and spin–orbit energies for NO is shown in Fig. 4.7. At its normal boiling point of 121°K one can see that Eq. (1.27) would have to include a number of energy levels that are close to kT in energy.

The calculation of χ_M for gaseous NO can be greatly simplified by the use of Van Vleck's concept of "low-frequency" elements. Since the rotational spacings are small compared to kT they all form "low-frequency" elements in Eq. (1.31*b*). However, the spacing between the $^2\Pi_{3/2}$ and $^2\Pi_{1/2}$ levels is a "medium frequency," and so Eq. (1.31*b*) must be modified to account for this fact. The result of this calculation will be similar to the atomic $^2P_{3/2}$ and $^2P_{1/2}$ terms treated in Section 3.1. We shall also use the concept of parallel and perpendicular moments contributing to the first and last terms, respectively, in Eq. (1.27).

For pure $^2\Pi_{3/2}$ and $^2\Pi_{1/2}$ states without rotation the angular momentum $\mathbf{\Omega}$, quantized along the molecular z axis, would precess around the spaced-fixed Z axis. Figure 3.1 must be modified so that

$$\begin{aligned}\mu_{\parallel}^2 &= \mu_B^2(L_z + g_0 S_z)^2 \\ &= \mu_B^2(\Lambda + g_0\Sigma)^2 \end{aligned} \tag{4.11a}$$

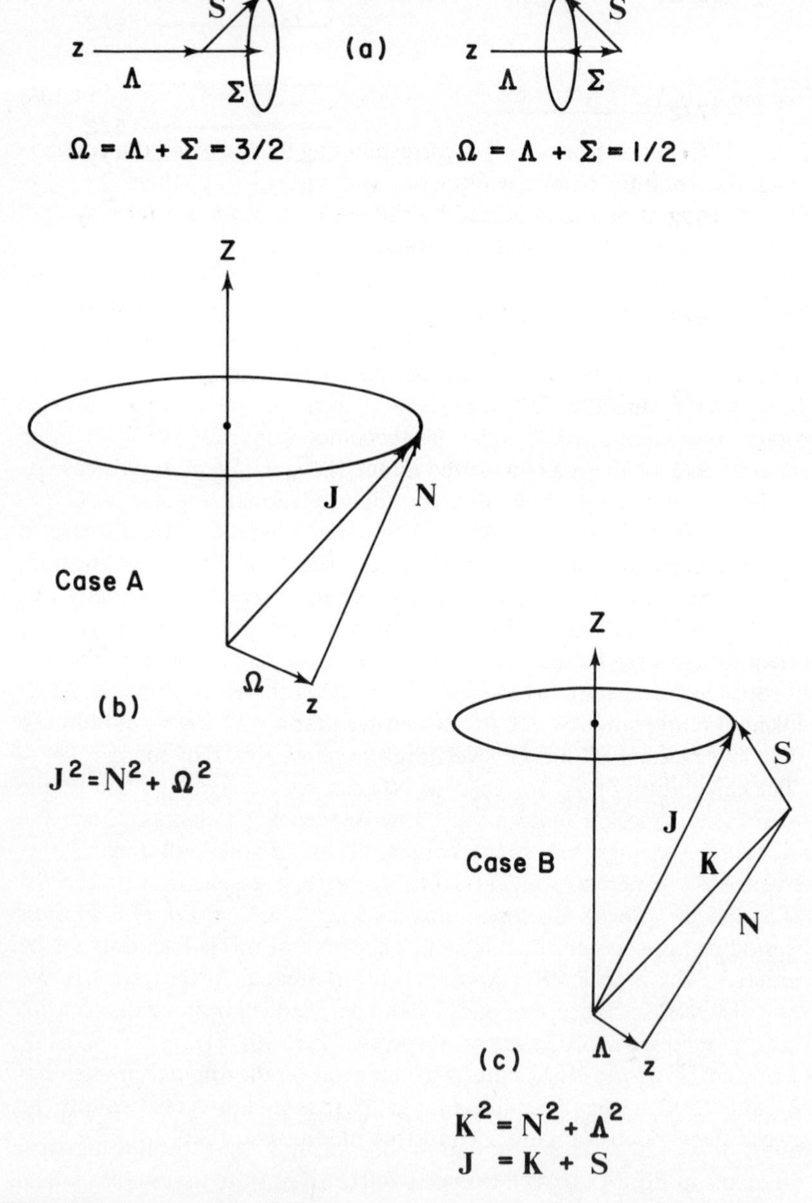

Figure 4.6. *Vector-model diagrams for the NO molecule. (a) Without rotational angular momentum. (b) and (c) With rotation angular momentum* **N**.

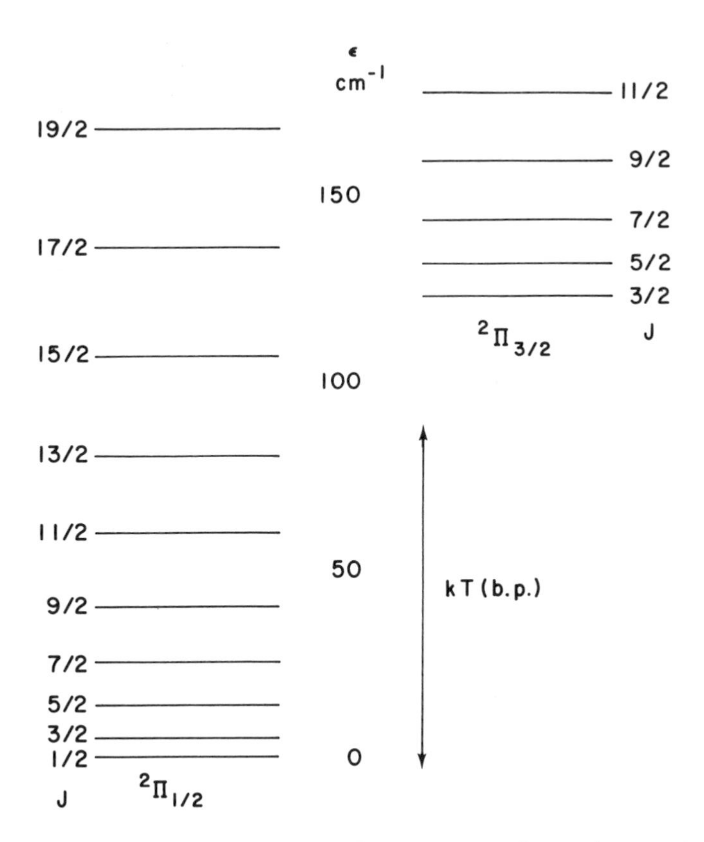

Figure 4.7. *Energy levels for NO including rotation with Hund's case A. The value of kT of the normal boiling point of* 121°K *is also indicated. The energy levels follow the equation* $\epsilon = B[J(J+1) - \Omega^2] + \lambda\Sigma\Lambda$ *where* $B = 0.80\ cm^{-1}$, *the rotational constant for* $^{14}N^{16}O$, *and* $\lambda = 124\ cm^{-1}$.

and

$$\begin{aligned}\mu_\perp^2 &= \mu_B^2 g_0^2(S_x^2 + S_y^2)\\ &= \mu_B^2 g_0^2(\mathbf{S}^2 - S_z^2)\\ &= \mu_B^2 g_0^2[S(S+1) - \Sigma^2].\end{aligned} \tag{4.11b}$$

With $g_0 = 2$ one has for the $^2\Pi_{3/2}$, $\mu_\parallel^2 = 4\mu_B^2$, for $^2\Pi_{1/2}$, $\mu_\parallel^2 = 0$ and for both, $\mu_\perp^2 = 2\mu_B^2$. The cancelation of the spin and orbital magnetic moments in Eq. (4.11*a*) is a property of the $^2\Pi_{1/2}$ state that makes it have no first-order paramagnetism.

These values for the parallel and perpendicular moments can be used in

Eq. (1.31*b*) and Eq. (3.20) if we take into account that both the $^2\Pi_{1/2}$ and $^2\Pi_{3/2}$ are populated. The rotational levels are "low-frequency" elements, but this rotational averaging eliminates the $2\Omega + 1$ degeneracy. As a result, from the two values for $\mu_{\|}$ one gets

$$\chi_M - N\alpha = \frac{N\mu_B^2}{3kT}\left[\frac{0 + 4e^{-\lambda/kT}}{1 + e^{-\lambda/kT}}\right] \tag{4.12}$$

and from $\mu_{\perp}$

$$\alpha(\text{para}) = \frac{4\mu_B^2}{3\lambda}\,\frac{1 - e^{-\lambda/kT}}{1 + e^{-\lambda/kT}}\,. \tag{4.13}$$

When these two equations are combined, the result has some similarity to Eq. (3.17). If the result is expressed in terms of μ_{eff}, we have

$$\mu_{\text{eff}} = 2\sqrt{\frac{1 - e^{-x} + xe^{-x}}{x + xe^{-x}}} \tag{4.14}$$

where $x = \lambda/kT$. A plot of Eq. (4.14) together with the experimental data for NO are shown in Fig. 4.8.

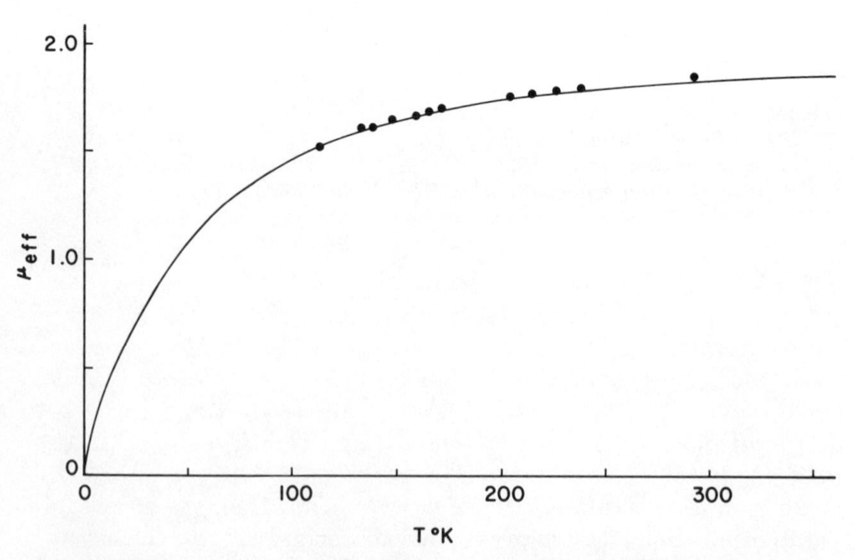

Figure 4.8. *Experimental and theoretical* μ_{eff} *for* NO(*g*). *Solid curve is Eq.* (4.14) *and the experimental points are taken from the reference given in Table* 1.2.

It can be seen from Fig. 4.8 that at high temperatures $\mu_{\text{eff}} \rightarrow 2$. The vector model can yield this same result, for at high temperatures the averaging of the $^2\Pi_{1/2}$ and $^2\Pi_{3/2}$ states corresponds to the uncoupling of $\mathbf{S}$ and $\mathbf{\Lambda}$ in the same way that $\mathbf{S}$ and $\mathbf{L}$ are uncoupled in atoms. With this uncoupling $\mathbf{S}$ and $\mathbf{\Lambda}$ would precess separately and

$$\mu_{\text{eff}}^2 = g_0^2 S(S+1) + \Lambda^2 \tag{4.15}$$

This gives $\mu_{\text{eff}} = 2$ for a $^2\Pi$ term for $g_0 = 2$, $S = \frac{1}{2}$, and $\Lambda = 1$.

The development of Eqs. (4.14) and (4.15) depends so much on the general principles that we have tried to illustrate earlier that we have not attempted to expose every step in their derivation. It is clear, however, that the direct use of Eq. (1.27) to the energy levels of Fig. 4.7 might possibly yield Eq. (4.14), but only at the expense of considerable algebra and a knowledge of additional quantum mechanics. Perhaps at this point we have at last concluded our explanation of the general equations derived in Section 1.4 for paramagnetic systems.

4.5. Molecular Orbital Theory for Transition-Metal Complexes

While the crystal field theory explains most of the features of the magnetic properties of transition-metal complexes, it is not a realistic theory, for covalent bonding must be involved. Ligands such as H_2O, NH_3, Cl^-, CO, and CN^- are not just negative charges that repel the *d* orbitals of a transition-metal ion. They are electron-rich ligands with at least one pair of electrons that can form bonds with the central ion.

Most chemical bonds can be classified as either a σ or a π bond. The σ bond has an electron density that has cylindrical symmetry, whereas the π bond has only a twofold symmetry. The famous double bond in ethylene consists of one σ bond and one π bond. In a diatomic molecule a π molecular orbital can hold four electrons and it forms the basis for two π bonds. In a polyatomic molecule one still uses the term π orbital, but because of the lower symmetry these π electrons have completely quenched orbital angular momentum and no orbital degeneracy. As a result the π orbitals in a polyatomic molecule can hold only two electrons.

In H_2O and NH_3 the only electrons that are free to bond to a transition-metal ion are the well-known "lone-pair" electrons and they can only form σ bonds. The other electrons are all too busy bonding to the protons. In Cl^- all the electrons are more or less equally free to form either

σ or π bonds, and the very strong complexes formed by CO and CN^- can only be explained if both σ and π bonds are assumed. The elements of the theory are nicely revealed if we restrict our ligands to σ bonds, and we shall make this assumption.

A σ orbital from a ligand and a σ orbital from the central ion can form one bonding σ molecular orbital and one antibonding σ^* molecular orbital. In an octahedral complex this requires each of the six ligands to have one σ orbital available for bonding and the central ion to also have a suitable set of six σ orbitals. In a transition-metal ion, symmetry considerations show that the d orbitals can form only two σ orbitals for bonding in an octahedral complex. These are the $d(z^2)$ and the $d(x^2 - y^2)$, which point directly at the ligands. To form the other four σ orbitals

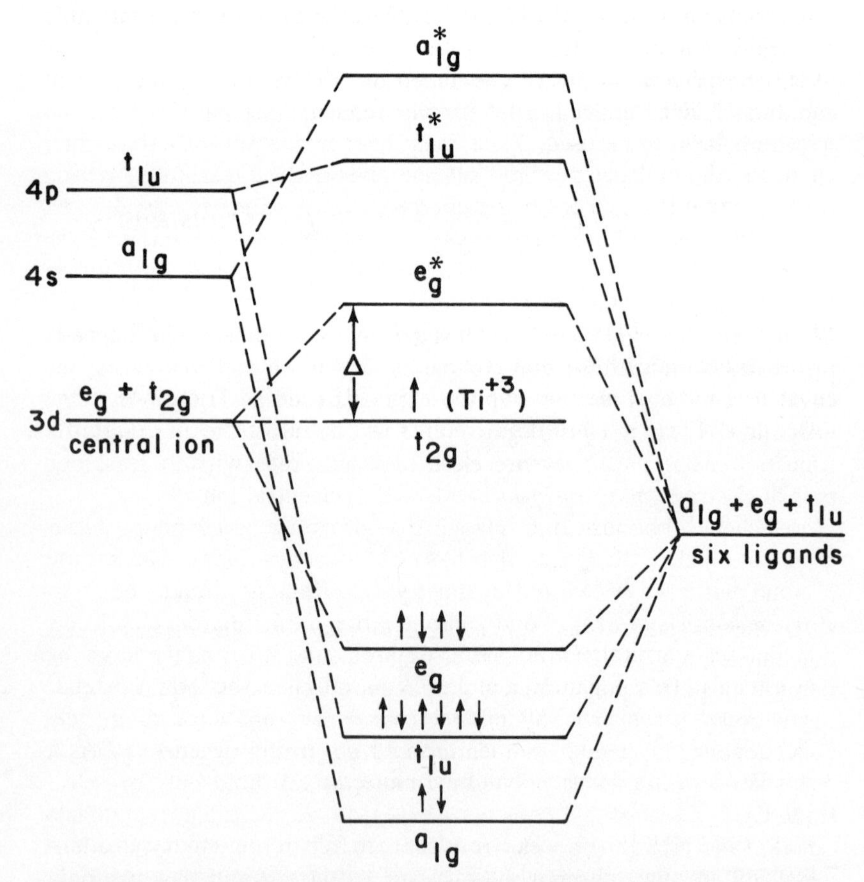

Figure 4.9. *Formation of molecular-orbital energy levels for octahedral transition-metal complexes with only sigma bonding. The symbols a_{1g}, e_g, t_{1u}, and t_{2g} are symmetry designations appropriate to octahedral symmetry (point group O_h).*

one must use the normally unoccupied 4*s* and 4*p* atomic orbitals of an iron-group transition-metal ion. A diagrammatic energy-level diagram for this bonding is shown in Fig. 4.9.

It can be seen from this figure that the six ligands donate a total of 12 electrons which occupy the three bonding sigma orbitals. Because of the high symmetry of an octahedral complex most of these orbitals have an orbital degeneracy such that they can hold more than one pair of electrons. In the σ-bonding approximation the lowest molecular orbital for the electrons from the central ion is a t_{2g} orbital, which is a pure atomic orbital, so that with only σ bonding, Ti^{+3} has its odd electron in the same orbital for both the crystal field and molecular-orbital approximations. However, in Fig. 4.9 one sees that Δ in the molecular-orbital scheme is determined by the position of an antibonding e_g^* energy level. The stronger the σ bonding, the larger Δ is, because the greater the energy is of the antibonding levels.

On the experimental side it is observed that the most obvious failure of the crystal field model is that greatly reduced spin–orbit interaction constants have to be used. Since most ligands contain only the lighter elements, the molecular-orbital scheme gives a good qualitative reason for this, but it is very hard to get any quantitative values for the reduced spin–orbit constants from the theory. As we shall see in Chapter 8 the molecular-orbital theory also explains qualitatively the observation in EPR of ligand hyperfine interactions.

When one includes π bonding, the nonbonding t_{2g} orbital also contains ligand contributions, so that whereas one may neglect π bonding because it is weak, it may be important in its ligand contributions to the lower crystal field orbital. These points will be illustrated in Chapter 8.

References

K. D. Bowers and J. Owen, *Paramagnetic Resonance II*, Reports in Progress in Physics, Vol. XVIII, pp. 304–373 (1955). Summary of EPR work on transition-metal, lanthanide, and actinide ions in crystals. Particularly clear with respect to magnetic axes in single crystals.

C. A. Coulson, *Valence*, Oxford University Press, Inc., New York (1961). A standard book on elementary molecular structure.

H. B. Gray, *Electrons and Chemical Bonding*, W. A. Benjamin, Inc., Menlo Park, Calif. (1964). An easy-to-read introduction to the molecular-orbital theory of bonding. Designed for first- and second-year undergraduates.

J. F. Harrison and L. C. Allen, *J. Amer. Chem. Soc.*, **91,** 807 (1969). One of the early calculations showing that CH_2 has a bent triplet ground state.

G. Herzberg, *Molecular Spectra and Molecular Structure*, Van Nostrand Reinhold Company, Vol. I: *Spectra of Diatomic Molecules* (1950) and Vol. III: *Electronic Spectra and Electronic Structure of Polyatomic Molecules* (1966). The classic books on molecular spectroscopy and energy levels.

J. W. Linnett, *The Electronic Structure of Molecules*, John Wiley & Sons, Inc., New York (1964). A new approach particularly suitable for free radicals.

Problems

4.1. Use Fig. 4.1 to predict the electron configurations and lowest energy terms for F atom, linear NH_2, and bent NH_2. Check your answers for NH_2 in Herzberg, Vol. III, p. 493 and Table 62, p. 583.

4.2. For $Cu(H_2O)_6^{+2}$ it has been found that $g_{\|} = 2.40$ and $g_{\perp} = 2.10$. Calculate the average g value that one would find from the susceptibility of a solution or powder. Compare this with that given in Problem 3.3.

4.3.* What concentration of free radical with $g = 2.00$ in benzene would just cancel the diamagnetism of this solvent at 20°C? The density of benzene is 0.879 g/ml. Express your answer in moles of free radical per liter of benzene.

4.4. The solubility of air in water at 20°C and 1 atm is such that 6.4 cm³ of O_2 dissolves in 1 liter of water. Determine the percentage of error introduced in the diamagnetic susceptibility of water at 20°C by dissolved O_2.

4.5. In single crystals of $CuSO_4 \cdot 5H_2O$ it has been observed that despite triclinic crystal symmetry that the susceptibility tensor is nearly axial with $\chi_M^1 = \chi_M^2 = 1255 \times 10^{-6}$ and $\chi_M^3 = 1532 \times 10^{-6}$ at 293°K.

(a) Show that these values are consistent with the powder value given in Table 1.1.

(b) It has been found that there are two Cu^{+2} ions per unit cell and each has a tetragonally distorted coordination with $g_{\|} = 2.40$ and $g_{\perp} = 2.07$. Show that the axial nature of the single crystal susceptibility is consistent with the z axes of these Cu^{+2} ions as being located at 82° with respect to each other. Use χ_M^{av} and Eq. (4.5*b*) to evaluate an average diamagnetic correction and determine theoretical values for χ_M^1, χ_M^2, and χ_M^3. Show that χ_M^1 and χ_M^2 should be very similar.

4.6.* Single crystals of $CuCl_2 \cdot 2H_2O$ are orthorhombic and it is found that at 290°K $\chi_M^1 = 1465$, $\chi_M^2 = 1250$, and $\chi_M^3 = 1600 \times 10^{-6}$. Show that, with only small diamagnetic corrections, these are consistent with EPR measurements where it is found that $g_a = 2.187$, $g_b = 2.037$, and $g_c = 2.252 \pm 0.005$. The crystal is reported to have two Cu^{+2} ions per unit cell but the published EPR spectrum has not fully resolved the difference between them.

4.7. What are the vector-model solutions to Eq. (4.8) when H is along z? In the vector model this corresponds to Z along the molecular z axis. Neglect any difference between $g_{\|}$ and $g_{\perp}$. Remember that for a triplet state $S = 1$ and $M_S = 0$ or ± 1. These are the levels plotted as solid lines in Fig. 4.4.

4.8. The triplet-state energy levels for Eq. (4.8) with $g_{\|} = g_{\perp} = g$ are given by $\epsilon = D/3$ and by

$$\epsilon = -\frac{D}{6} \pm \sqrt{\left(\frac{D}{2}\right)^2 + g^2\mu_B^2H^2}$$

when H is along x or y. These are the levels plotted as dashed lines in Fig. 4.4. Show that for large magnetic fields the correct vector-model solution to Eq, (4.8) with Z along x or y is

$$\epsilon = g\mu_B HM_S + D[\langle S_z^2\rangle_{av} - \tfrac{1}{3}S(S+1)]$$

where

$$\langle S_z^2\rangle_{av} = \tfrac{1}{2}[S(S+1) - M_S^2]$$

and $M_S = 0$ or ± 1. In the vector model with Z along x,

$$\langle S_y^2\rangle_{av} = \langle S_z^2\rangle_{av},$$

and, of course,

$$\langle S_x^2\rangle_{av} + \langle S_y^2\rangle_{av} + \langle S_z^2\rangle_{av} = S(S+1).$$

4.9. If H is along x for Eq. (4.7), show from the vector model that for large magnetic fields the energy levels are given by

$$\epsilon = \frac{D}{3} - E \qquad \text{and} \qquad \epsilon = -\frac{D}{6} + \frac{E}{2} \pm g\mu_B H.$$

Use the results of Problem 4.8 for $\langle S_z^2\rangle_{av}$. In this case

$$\langle S_x^2 - S_y^2\rangle_{av} = \langle S_x^2\rangle_{av} - \langle S_y^2\rangle_{av}$$

since S_x is quantized in the vector model with H along x. Neglect any difference between the three g values.

Solutions:

4.3. From Table 1.2 for benzene at 20°C,

$$\chi_M = -54.85 \times 10^{-6},$$

or for 1 liter,

$$\chi = -54.85 \times \frac{879}{78} \times 10^{-6}$$
$$= -616 \times 10^{-6}.$$

With Eq. (4.4) and a solution in which $\chi(\text{net}) = 0$,

$$N = \frac{4 \times 1.38 \times 10^{-16} \times 293 \times 616 \times 10^{-6}}{4 \times (9.27 \times 10^{-21})^2}$$

$$= 2.9 \times 10^{23} \text{ atoms.}$$

This corresponds to

$$n = \frac{2.9 \times 10^{23}}{6.0 \times 10^{23}} = 0.48 \text{ mole/liter.}$$

4.6. From Eq. (4.4)

$$\chi_M^i = g_i^2 \frac{6.023 \times 10^{23}(0.273 \times 10^{-21})^2}{4 \times 1.381 \times 10^{-16} \times 290}$$

$$= g_i^2 \times 323.3 \times 10^{-6}.$$

This gives

	$a(1)$	$b(2)$	$c(3)$
χ(para) =	1546	1342	1640
χ(obs) =	1465	1250	1600
χ(dia) =	81	92	40

All values are $\times 10^{-6}$.

5

Molecular Diamagnetism

IN THIS CHAPTER we shall be concerned with those molecules that have no permanent electronic magnetic moment. A necessary condition for the electronic spin moment to be zero is that the molecules have an even number of electrons. This is not very restrictive, for most of the common stable molecules, such as H_2, CH_4, and SO_2, have an even number of electrons. Although these molecules usually form the basis for diamagnetic substances, we cannot always neglect the origins of paramagnetism even for these materials. But before we attack the topic of molecules, we must consider simple diamagnetic atoms and the origins of diamagnetism.

5.1. Diamagnetic Atoms and Larmor Precession

Those atoms with an even number of electrons and with closed orbital configurations such as $(1s)^2$, $(1s)^2(2s)^2$ and $(1s)^2(2s)^2(2p)^6$ are examples of ^1S ground states where $S = 0$, $L = 0$, and $J = 0$. The paramagnetic Hamiltonian, Eq. (1.19), contributes nothing to the energy of such states. This can be seen on a quantum-mechanical basis from Eq. (3.8) for $\epsilon_i^{(1)} = 0$, since with $J = 0$ and $S = 0$, M_J and M_S must also be zero. In addition, it can be shown that $\epsilon_i^{(2)} = 0$. For an atom, $\epsilon_i^{(2)}$ depends upon electron spin paramagnetism by Eq. (3.9). The matrix elements in this equation can also be related to M_S^2 if we use the principles of matrix multiplication, and we find that

$$\begin{aligned} \langle\psi_i^0|\, \mathsf{S}_Z^2 \,|\psi_i^0\rangle &= \sum_j \langle\psi_i^0|\, \mathsf{S}_Z \,|\psi_j^0\rangle\langle\psi_j^0|\, \mathsf{S}_Z \,|\psi_i^0\rangle \\ &= \sum_j |\langle\psi_i^0|\, \mathsf{S}_Z \,|\psi_j^0\rangle|^2 \qquad (5.1a) \\ &= M_S^2. \qquad (5.1b) \end{aligned}$$

The matrix elements of Eq. (5.1a) are those in Eq. (3.9) that yield $\epsilon_i^{(2)}$ for an atom. However, from Eq. (5.1b) with a singlet state, $S = 0$ and $M_S = 0$. Since it is necessary that each matrix element in Eq. (5.1a) be zero, $\epsilon_i^{(2)}$ must also be zero. These facts can be summarized by saying that in singlet states there is no electron spin contribution to paramagnetism, and in an atom with $L = 0$ there is also no orbital contribution to paramagnetism. Another proof can be based upon the vector-model diagram for atoms in Fig. 3.1. When $S = 0$, $L = 0$, and $J = 0$ there is neither a parallel nor a perpendicular component of the magnetic moment. This correctly predicts that both $\epsilon^{(1)}$ and $\epsilon^{(2)}$ are zero for such atoms. As a result of these considerations we can conclude that the ^1S ground state in closed-shell atoms should give no paramagnetic contributions to χ_M since all its terms in Eq. (1.27) are zero. Since these atoms are diamagnetic, we must now discover the origins of diamagnetism.

The term in the Hamiltonian for diamagnetism is derived in Section A.4. We shall discuss this derivation later in this section. At this moment, let us accept without proof that this term is

$$\mathsf{H}'(\text{dia}) = \frac{e^2H^2}{8mc^2}(\mathsf{X}^2 + \mathsf{Y}^2) \qquad (5.2)$$

where it must be summed over all the electrons. The most important part of the energy that results from Eq. (5.2) can be written very simply in terms of the average value of $(X^2 + Y^2)$ over the wave functions for the ground state of the molecule and

$$\begin{aligned} \epsilon_i(\text{dia}) &= \frac{e^2H^2}{8mc^2}\langle\psi_i^0|\, X^2 + Y^2 \,|\psi_i^0\rangle \\ &= \frac{e^2H^2}{8mc^2}\langle X^2 + Y^2\rangle_i \qquad (5.3) \end{aligned}$$

where these equations must always be summed over all the electrons. For an atom with spherical symmetry

$$\langle X^2\rangle = \langle Y^2\rangle = \langle Z^2\rangle = \tfrac{1}{3}\langle r^2\rangle,$$

and so

$$\epsilon_i(\text{dia}) = \frac{e^2H^2}{12mc^2}\langle r^2\rangle_i. \tag{5.4}$$

LARMOR FREQUENCY

For atoms it is quite easy to see that diamagnetism is a result of the force generated between a rotating electron and a magnetic field. In Fig. 5.1 it is assumed that we have a classical atom with a single electron rotating with increasing ϕ in the XY plane. In the absence of the field the electron will be attracted to the nucleus by a Coulombic force, and balanced with this force is the outward centrifugal force, where

$$\begin{aligned} f(\text{centrifugal}) &= \frac{mv^2}{r} \\ &= m\omega^2 r. \end{aligned} \tag{5.5}$$

In the presence of the magnetic field H along the Z axis there is an additional force,

$$\begin{aligned} \mathbf{f}(\text{magnetic}) &= \frac{q}{c}\,\mathbf{v}\times\mathbf{H} = \frac{q}{c}(\boldsymbol{\omega}\cdot\mathbf{H})\mathbf{r} \\ &= -\frac{e\omega H}{c}\,\mathbf{r}. \end{aligned} \tag{5.6}$$

This force is along r, but because electrons are negatively charged, it opposes the centrifugal force for the direction of rotation indicated in Fig. 5.1. If we judge the effect of the magnetic field from the viewpoint of the field-free atom, ω must increase to compensate for the magnetic force. This increase in ω can be evaluated by equating the magnetic force with the increase in centrifugal force, so that

$$\begin{aligned} \frac{er\omega H}{c} &= \delta(m\omega^2 r) \\ &= 2mr\omega\,\delta\omega, \end{aligned} \tag{5.7a}$$

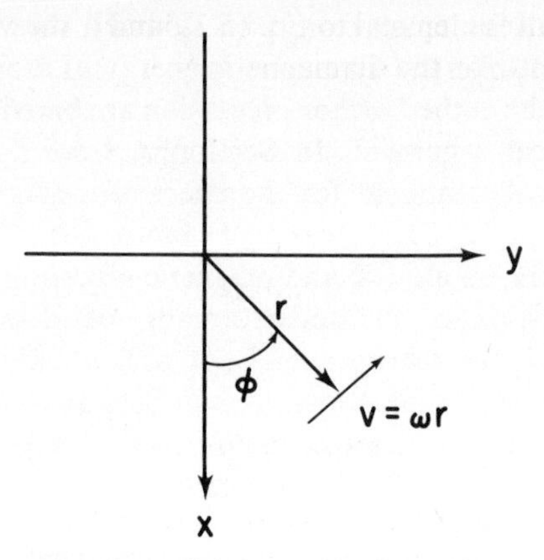

Figure 5.1. *Direction of positive rotation in the XY plane. The magnetic field is taken as pointing along Z out of the plane of the paper.*

or the increase in ω is

$$\delta\omega = \omega(\text{Larmor}) = \frac{eH}{2mc}. \tag{5.7b}$$

If in Fig. 5.1 we assumed that the electron was rotating in the negative ω direction, that electron would have to decrease its angular frequency to compensate for the force from Eq. (5.6). The result for both possible directions of rotation is a net precession of the total electronic motion corresponding to positive ω values and at a frequency given by Eq. (5.7*b*). This observation was first made by J. Larmor in his early work *Aether and Matter* (1900), and the angular frequency given by Eq. (5.7*b*) is called the Larmor frequency.

The effect of this Larmor precession of all the electrons in a classical atom is to generate an induced magnetic moment. According to Eq. (1.3), since electrons are negatively charged, this induced moment opposes the field, and if we combine Eq. (1.3) with (5.7*b*),

$$\mu(\text{Larmor}) = -\frac{e^2H}{4mc^2}(X^2 + Y^2). \tag{5.8}$$

From Eqs. (1.20) and (1.21) one can see that the energy due to an induced moment is $\epsilon = -\frac{1}{2}H\mu(\text{induced})$. If we also average over the rotational motion of all the electrons,

$$\epsilon = \frac{e^2H^2}{8mc^2}(X^2 + Y^2)_{\text{av}}. \tag{5.9}$$

This classical result is identical to Eq. (5.3), and it shows that the Larmor precession accounts for the diamagnetic energy of atoms.

These relations about the Larmor precession are based upon a simplified classical-mechanical argument. In Section A.4 we show the complete classical-mechanical treatment for the effect of a magnetic field upon a moving charged body. This treatment appears in Chapter I of Van Vleck's classic work on electric and magnetic susceptibilities and Section A.4 is fundamental to an understanding of diamagnetism. Many students may find this theory somewhat difficult and we shall try to explain the significance of the equations in Section A.4.

The heart of this derivation is contained in Eq. (A.10*b*). For a moving electron this becomes

$$p_X = mv_X - \frac{e}{c} A_X, \tag{5.10}$$

where there are similar equations for p_Y and p_Z. Since we are primarily concerned with angular momentum, we can use Eq. (A.16*b*) and obtain for the Z component of orbital angular momentum

$$P_Z = (Xmv_Y - Ymv_X) - \frac{e}{c}(XA_Y - YA_X). \tag{5.11a}$$

If we apply this equation to the problem of a single electron rotating in the XY plane, as shown in Fig. 5.1, we obtain

$$P_Z = mr^2\omega - \frac{e}{c}(XA_Y - YA_X). \tag{5.11b}$$

Our previous result for a magnetic field along the Z axis was that ω would increase in value by $eH/2mc$. As a result we would expect that P_Z would also be increased unless the second term in Eq. (5.11*b*) exactly canceled this increase. If P_Z does increase, we would find that H′(para) in Eq. (1.19) would have to include this change, but if P_Z does not change, then H′(para) would not be affected by the Larmor precession. In atoms it is possible to have both terms in Eq. (5.11*a*) exactly cancel, so that the Larmor precession does not affect H′(para). The obvious problem is the one of evaluating the second term in Eq. (5.11*b*).

CHOICE OF GAUGE

The components of the vector potential A_X, A_Y, and A_Z must satisfy the relation that $\mathbf{H} = \text{curl}\,\mathbf{A}$. This is not a complete prescription, for there are many possible vector potentials that will satisfy this relation even when H is taken as along the Z axis. The arbitrary choice of a set of components for **A** for a given **H** is known as the gauge condition. The gauge condition in Section A.4 is

$$A_Z = 0 \qquad A_Y = \frac{X}{2} H \qquad A_X = -\frac{Y}{2} H. \tag{5.12}$$

With this gauge one can see that P_Z is constant in Eq. (5.11*b*) if ω is increased by $eH/2mc$. The second term exactly cancels this increase in the first term if the gauge of Eq. (5.12) is used. For an atom it is possible to use a gauge other than Eq. (5.12), but then the Larmor precession will affect both the paramagnetic and diamagnetic terms in the resulting Hamiltonian.

In molecules electrons are not rotating around a single nucleus. As a result, the Larmor precession will have a more complicated form. It is no longer true that both terms in Eq. (5.11*a*) which depend upon the magnetic field exactly cancel. For this reason all molecules have nonzero contributions both from H′(para) and H′(dia). In the final analysis the division of the Hamiltonian between paramagnetic and diamagnetic parts depends entirely on the arbitrary gauge of Eq. (5.12). This gauge is unique for atoms but not for molecules. For molecules with paired electrons it is possible to find a general gauge that will reduce the "paramagnetic" contributions to χ_M to a minimum value but not to a zero value. As a result, χ_M for molecules will always contain contributions from "paramagnetism" as defined by Eq. (1.19) and from "diamagnetism" as defined by Eq. (5.2). The division is arbitrary, but the value of χ_M is independent of the choice of gauge that establishes this division.

Since the electron spin does not arise from the classical rotation of a charge the spin operator in Eq. (1.19) is not field dependent. For this reason the electron spin can only contribute to paramagnetism and we need not worry about spin diamagnetism.

EXPERIMENT AND THEORY

From Eq. (5.4) one can easily see that diamagnetic atoms have susceptibilities given by

$$\chi_M = -\frac{Ne^2}{6mc^2} \sum_i \langle r^2(i) \rangle \tag{5.13}$$

when, as is almost always the case, excited states have energies very much larger than kT. This is a very simple equation and it clearly shows that big atoms with many electrons should have the largest absolute values for their diamagnetic susceptibilities.

The smallest diamagnetic atom is gaseous helium. For this simple two-electron atom the calculated susceptibilities can be very much more accurate than the experimental ones, and there are at present two accepted experimental values. The older value is $\chi_M = -1.87 \times 10^{-6}$ and the new value is $\chi_M = -1.93_3 \times 10^{-6}$. These two values can be compared to theory with Eq. (5.13) and they yield $\langle r^2(i)\rangle = 1.19$ and 1.22_0, respectively, all in units of the Bohr radius a_0^2. The Hartree self-consistent field value is $\langle r^2(i)\rangle = 1.19$, in excellent agreement with the older experimental value. At one time it was thought that Hartree's value was in error, but high-speed digital computations that give very accurate energies have been done on helium. The most accurate of these computations shows that $\langle r^2(i)\rangle = 1.19348299$. This confirms the Hartree value for helium and clearly establishes the accuracy of theory for such a small atom. The limiting factor in the calculation of χ_M for helium is now the accuracy of the physical constants. For atoms as complex as argon the experimental values are still probably just as accurate as the presently published theoretical ones. The Hartree–Fock self-consistent field wave functions could only be expected to come within about 1 per cent of the experimental value for argon of $\chi_M = -19.6 \times 10^{-6}$.

Certain salts, such as NaCl and KF, are often considered to be composed of spherical closed-shell ions. If the diamagnetism of such salts is taken as the sum of the anion and cation susceptibilities, then by taking differences the validity of these assumptions can be tested. On the whole, fairly consistent differences are obtained. This does not necessarily mean that such salts are truly ionic, for a glance at Eq. (5.13) shows that for Cl^-, for example, the $1s$, $2s$, $2p$, and particularly $3s$ electrons all contribute to the susceptibility. The result is that small changes in the $3p$ electrons in Cl^- may not produce significant variations in χ_M. One can conclude that for most ionic solids χ_M can be readily approximated using simple ionic models together with a consistent set of ionic susceptibilities. Several sets of ionic susceptibilities have been tabulated and can be found in a number of places; see, for example, Table 1.2.

LAMB EQUATION

The magnetic field produced at the nucleus is of particular importance. This field can be measured quite accurately for several nuclei by means of nuclear magnetic resonance spectroscopy, and for a diamagnetic atom it can be calculated in a very direct way.

The magnetic field along the Z axis of a current loop parallel to the XY plane is given by

$$H(\text{loop}) = \frac{2\pi i(X^2 + Y^2)}{(X^2 + Y^2 + Z^2)^{3/2}} = \frac{2\mu}{(X^2 + Y^2 + Z^2)^{3/2}}. \tag{5.14}$$

We previously saw in Eq. (5.8) that a diamagnetic atom has an induced dipole moment that opposes the applied field. This means that the loop field will also oppose the applied field. If H_0 is the applied field and H is the actual field at the nucleus, we can define the diamagnetic shielding parameter σ by

$$H = H_0(1 - \sigma). \tag{5.15}$$

The σ values are very small and they are usually expressed in parts per million.

If we combine Eqs. (5.8) and (5.14), remembering that H is very close to H_0, we obtain for a nucleus in the XY plane

$$\sigma = \frac{e^2}{2mc^2} \frac{X^2 + Y^2}{r^3}. \tag{5.16}$$

This expression must be summed over all the electrons, and for a spherical atom we obtain for the quantum-mechanical average

$$\sigma = \frac{e^2}{3mc^2} \sum_i \langle 1/r(i) \rangle. \tag{5.17}$$

This equation was first derived by W. Lamb (1941).

The potential energy of attraction between the nucleus and the electrons is also proportional to $\langle 1/r \rangle$, so that self-consistent field wave functions do a rather good job for the calculation of σ values from the Lamb equation. The differences between the equations for σ and χ_M are quite fundamental and it is not possible to calculate σ from an experimental χ_M value. The "inner" electrons in an atom are more important in the σ equation and the "outer" electrons are more important for χ_M.

5.2. Diamagnetic Molecules

In atoms both the paramagnetic terms given by Eq. (1.19) could be exactly zero, but only the spin term can be zero for molecules. Atoms can be spherical and possess no orbital moment. In molecules we can only consider the orbital moment to be quenched so that it does not contribute to $\epsilon^{(1)}$, but $\epsilon^{(2)}$ is never zero.† The orbital moment in molecules must always contribute to $\epsilon^{(2)}$, and so molecules must possess a contribution to χ_M which is due to second-order paramagnetism. If we use Eq. (1.29) for a molecule in a singlet state, we obtain

$$\chi_M = -\frac{Ne^2}{4mc^2}\sum_i \langle X^2(i) + Y^2(i)\rangle_{\text{av}} + 2N\mu_B^2 \sum_{n\neq 0} \frac{|\langle\psi_0^0|\,\mathsf{L}_Z\,|\psi_n^0\rangle|_{\text{av}}^2}{\epsilon_n^0 - \epsilon_0^0}, \tag{5.18}$$

where the first term arises from the diamagnetic part of the Hamiltonian and the second from the paramagnetic part. Since most diamagnetic molecules do not have low-lying excited states, $(\epsilon_n^0 - \epsilon_0^0)$ is assumed to be much larger than kT. As a result, the second term is one of Van Vleck's "high-frequency" terms and is independent of temperature.

Molecules are not spherical, and so both terms in Eq. (5.18) must be averaged over all molecular orientations. The first term can also be written like Eq. (5.13). An interesting feature of Eq. (5.18) is that while the Z axis is defined as along the applied field, the center of the coordinate system is not defined in this equation. One could, for example, select a different coordinate system for each electron, but in this case the second term would also have to be summed over each electron and cross terms would also be formed by the second term. The fact that Eq. (5.18) has no defined center for its coordinate system can be used to advantage in certain calculations. However, it is also clear that when the coordinate system is located away from the center of motion of the electrons, both terms become larger and the exact value of χ_M depends more on the difference between two large numbers.

One of the limitations of Eq. (5.18) is that it is based upon quantum-mechanical perturbation theory. In this theory, as we can see, the paramagnetic term depends upon the properties of the excited states. If one uses the quantum-mechanical variational principle, this dependence upon excited states can be avoided. The variational principle is based upon the fact that the correct wave function for a molecule, even in a magnetic field, is the one that gives it the lowest value of $\langle\mathsf{H}\rangle$. The key is to find a good field-dependent wave function that can be varied until $\langle\mathsf{H}\rangle$ is a minimum. This method was first applied to molecules in magnetic

† Van Vleck showed, after some debate, that this is true. His argument uses Eq. (5.1*a*) written for L_Z. Since $\mathbf{L}_Z$ is not quantized in a molecule, except under very special circumstances, then $\langle\mathbf{L}_Z^2\rangle$ is not exactly zero and the matrix elements for orbital angular momentum are also not zero for molecules.

fields by J. Tillieu and G. Guy (1954) and it has been extended by other workers. It is an interesting method but we shall not illustrate its use in detail.

Since the second term in Eq. (5.18) depends upon all the excited states of a molecule, it is very difficult to determine theoretically. As we will show later, the rotational magnetic moment of a diamagnetic molecule also depends upon these same matrix elements. From a measurement of the rotational magnetic moment it is now possible to make an independent determination of the value of the second term. For H_2, for example, the second term in Eq. (5.18) is found to be equal to $+0.085 \times 10^{-6}$, where the coordinate system is located at the center of mass of the molecule. The χ_M value is the sum of both terms and it has been measured to be -3.99×10^{-6}. One can see that for H_2, χ_M(dia) is much more important than χ_M(para) in determining the net value of χ_M.

If χ_M(para) was always as small as it is in H_2, then it could be almost ignored, but in most molecules it appears to be at least 10 per cent of χ_M and in some cases as large in value as χ_M itself. A large number of rotational magnetic moments are now available based upon the Zeeman effect in rotational spectroscopy. A few of these results are given in Table 5.3 and the theory for linear molecules is discussed in Section 5.3. With such values one can ignore the theoretical problem of evaluating the second term in Eq. (5.18) and concentrate on the first term.

A number of calculations have been done for the $\langle r^2 \rangle$ in CH_4. Some of these results are given in Table 5.1. Four sets of wave functions are compared and the Pitzer set results from a detailed self-consistent-field calculation with optimized parameters to give the lowest energy. Since CH_4 has a neonlike structure the atomic $1s$, $2s$, and $2p$ orbitals become

Table 5.1. *Calculated χ_M for Methane*[a]

Wave function	$\langle r^2 \rangle$[b]			χ_M[c]
	$1a_1$	$2a_1$	$1t_2$	
Coulson (1942)	0.026	0.4276	1.405	-17.1×10^{-6}
Woznick (1964)	0.027	1.005	1.347	-19.4×10^{-6}
Palke–Lipscomb (1966)	0.027	1.115	1.277	-18.8×10^{-6}
Pitzer (1967)	0.027	1.098	1.267	-18.6×10^{-6}

[a] Taken from R. Hegstrom and W. N. Lipscomb, *J. Chem. Phys.*, **46**, 4538 (1967).

[b] In units of 10^{-16} cm² per electron.

[c] Calculated using the first term in Eq. (5.12) for χ_M(dia) with χ_M(para) = 9.29×10^{-6} as determined from the rotational magnetic moment of CH_4.

$1a_1$, $2a_1$, and $1t_2$ molecular orbitals in the tetrahedral symmetry. The theoretical values for $\langle r^2 \rangle$ were used to calculate the first term in Eq. (5.12). The second term in this equation was calculated from the experimental value for the rotational magnetic moment and from this it was determined that $\chi_M(\text{para}) = 9.29 \times 10^{-6}$. The χ_M values in Table 5.1 represent the sum of both terms. Even the fairly old Coulson wave functions give a reasonable value for $\chi_M(\text{dia})$, so that all the χ_M values differ little more than 10 per cent. For CH_4 it can also be seen that $\chi_M(\text{para})$ is about 50 per cent of the absolute value of χ_M. The two best experimental values of χ_M are -12.2 and -17.4×10^{-6}. The second value is the more recent determination, and the calculated values shown are clearly as accurate as any direct experimental χ_M for CH_4.

A purely empirical method of estimating the diamagnetic susceptibilities of molecules was formulated by P. Pascal (1910). He developed a series of constants for each atom, including bonding effects, which when added can often give a reasonable estimate of χ_M values for diamagnetic molecules. There is little or no theory to this method and a tabulation of Pascal's constants can be found in most reference books.

DIAMAGNETIC ANISOTROPY

If single crystals are examined in a Faraday balance, it is possible to determine the three principal values of the susceptibility tensor. As it was for a paramagnetic substance, if the crystals are cubic, tetragonal, orthorhombic, or hexagonal, these principal magnetic axes must lie along the crystal *a*, *b*, and *c* axes, but with lower symmetry there may be no correspondence between the two axis systems. It is not easy to quantitatively explain diamagnetic anisotropy, particularly since it depends upon the orbital paramagnetism, but aromatic systems do form a class of compounds where this anisotropy is large and easily explained.

In aromatic systems the conjugated π electrons are highly delocalized. One finds that $-\chi_M$ is particularly large when the magnetic field is oriented perpendicular to the plane of conjugation. This direction will define the molecular-fixed z axis. These π electrons can be considered as occupying p_z orbitals on the carbon atoms of a benzene molecule. The delocalization means that these π electrons can jump relatively easily from carbon atom to carbon atom and in benzene their potential function can be approximated by one with essentially cylindrical symmetry. L. Pauling (1936) pointed out that with the approximation of a cylindrical potential function the orbital contribution due to filled π orbitals is zero.

The zero orbital contribution in this case can be seen from Eq. (5.1) if we write this equation for orbital angular momentum along the molecular z axis. In diatomic molecules with, of course, exact cylindrical symmetry we have seen that L_z can be quantized. This forms the basis of our designations Σ, Π, Δ, etc., for the electronic states of diatomic molecules. In the benzene molecule the six π electrons form what is equivalent to a Σ term, if we use the approximation of a cylindrical potential function. This assumes that half the electrons are rotating clockwise and half are rotating counterclockwise. If we rewrite Eq. (5.1) for the molecular z component of orbital angular momentum, we obtain $\langle L_z^2 \rangle = \Lambda^2$. In a Σ term one only has $\Lambda = 0$. As a result, all the matrix elements for second-order paramagnetism must vanish when the field is along the axis of a cylindrically symmetric potential function. For benzene these second-order paramagnetic terms may not be exactly zero, but they are at least very small. If we then take Eq. (5.2) for a cylindrical axis system orientated along the z axis where $\langle X^2 \rangle = \langle Y^2 \rangle = \frac{1}{2}\langle \rho^2 \rangle$, then

$$\chi_M^z(\pi) = -\frac{Ne^2}{4mc^2} \sum_i \langle \rho^2(i) \rangle \tag{5.19}$$

where ρ is the radial distance of the π electrons from the z axis.

Figure 5.2 shows the crystal structures of graphite and benzene. In graphite the π electrons form a series of large planar networks that are perpendicular to the axis of a hexagonal crystal system. In a hexagonal system the a and b axes must have equal susceptibilities and it is found that $\chi_M^c = \chi_M^z = -274 \times 10^{-6}$ and $\chi_M^a = \chi_M^b = \chi_M^x = \chi_M^y = -4.8 \times 10^{-6}$. If we use Eq. (5.19) and 5×10^{-6} to correct for the susceptibility due to the other electrons, then with one electron per mole of carbon we find that $\langle \rho^2 \rangle^{1/2} = 8.0$ Å. For comparison, the distance from the center of the benzene ring to each carbon is equal to the C—C bond and is 1.39 Å. Although there is no well-defined radius for a cylindrical potential function in graphite, the size of the calculated radius is very impressive.

A more detailed theory has been developed by F. London (1937). His theory is based upon the Hückel theory for π electrons in conjugated hydrocarbons. For molecules like benzene and naphthalene London's theory predicts anisotropies close to those given by Pauling's method. For conjugated hydrocarbons that do not have 6, 10, or 14 π electrons (the $4n + 2$ series) London's theory indicates a more complicated behavior. Such π systems possess an orbital degeneracy and for these cases the orbital contribution to χ_M cannot be neglected. Pauling's theory is usually spoken of as a ring-current model and it is common to refer to ring currents as the source of the anisotropy.

J. I. Musher (1965) has tried to discredit the ring-current model by pointing out that Eq. (5.19) is nothing more than the sum of $\langle X^2 + Y^2 \rangle$

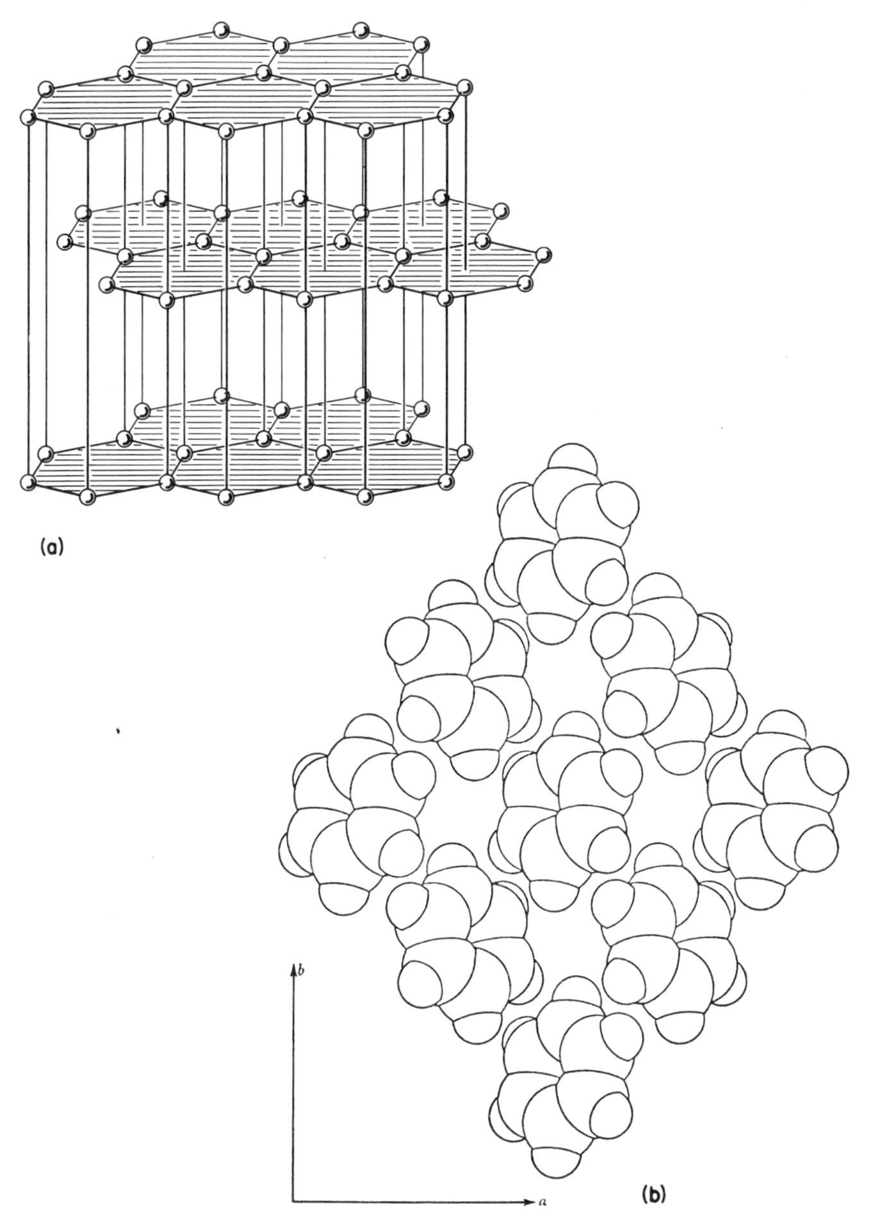

Figure 5.2. (*a*) *Crystal structure of graphite.* (*b*) *Crystal structure of benzene as viewed down the c axis.* [*After E. G. Cox, D. W. J. Cruickshank, and J. A. S. Smith, Proc. Roy. Soc.* (*London*), *A*, **247**, *1* (*1958*).] (*Note: The crystal contains another type of layer on top of this one in which the molecules are staggered as in graphite but twisted in the opposite sense. There are four atoms in a unit cell, two in this layer and two in the next.*)

over the individual atoms. The reason that this sum is large is that aromatic hydrocarbons are large planar molecules. The important point in Pauling's theory is not that $\langle X^2 + Y^2 \rangle$ is large, but that the second-order paramagnetism is small. Highly delocalized π electrons are assumed to be able to move freely in the rings and it is this movement in a cylindrical potential function that removes the second-order paramagnetism. Davies prefers to substitute delocalization susceptibility for ring current in his theoretical discussions.

Table 5.2 shows that the anisotropies for aromatic hydrocarbons increase with increasing number of benzene rings, but the values for the chlorobenzenes also show that the σ bonding must contribute to the

Table 5.2. *Anisotropic Susceptibilities from Crystal Measurement*[a]

Molecule (axes y, x)	**cm³ mole⁻¹ × 10⁻⁶**		
	χ_M^x	χ_M^y	χ_M^z
	−34.9	−34.9	−94.6
Cl, Cl	−78.3	−50.3	−120.2
Cl, Cl, Cl, Cl, Cl, Cl	−128.0	−128.0	−182.0
	−54.7	−52.6	−173.5
	−75.8	−62.6	−251.6
	−80.6	−80.6	−303.0

[a] Taken from A. A. Bothner-By and J. A. Pople, *Annual Review of Physical Chemistry*, **16,** 43 (1965).

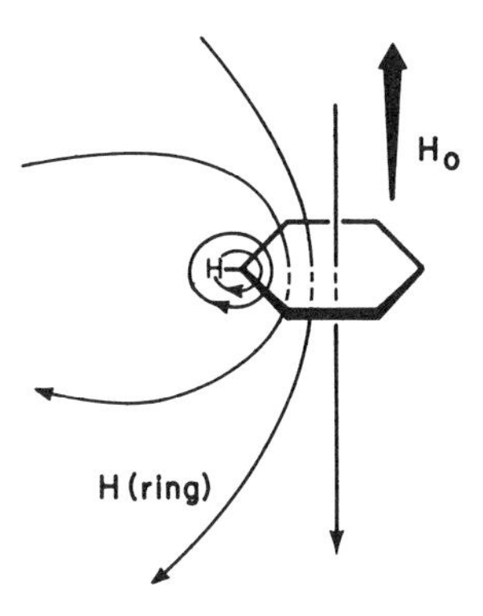

Figure 5.3. *Ring-current field in benzene resulting from the Larmor precession of its π electrons.*

anisotropy. Some of the present-day theories ascribe a fair fraction of the anisotropy to σ as well as π bonding.

A ring current in benzene would introduce a magnetic field as illustrated in Fig. 5.3. With NMR it is possible to measure the local field both inside and outside benzene rings. The ring-current model appears to be quite able to make semiquantitative predictions that agree with the NMR results. Many theoretical problems remain to be solved in the area of diamagnetic anisotropy, and the present literature is quite contradictory. Interested readers are encouraged to read only the more recent reviews such as to be found in Davies' work and in the article cited in Table 5.2.

5.3. Rotational Magnetic Moments

In our previous discussions in this chapter we have assumed that susceptibility measurements were being made on solids or liquids where the rotation of molecules can be neglected. However, when spectroscopic measurements are made on gases, an analysis of the effects of rotation becomes an essential part of a study of molecules in magnetic fields. To simplify our equations we shall restrict our analysis to linear molecules. The Hamiltonian for the rotational energy of a linear $^1\Sigma$ molecule can be written as

$$\mathsf{H}(\text{rot}) = \frac{h^2\mathbf{J}^2}{8\pi^2 I} = B\mathbf{J}^2 \qquad (5.20)$$

where I is the moment of inertia of the linear molecule perpendicular to its symmetry axis and B is called the rotational constant. In this equation

J is the total angular momentum of the molecule. Since it is a $^1\Sigma$ molecule, **J** is primarily the angular momentum of the rotating nuclei with only a very small contribution from the electrons. As we learned earlier, the square of the total angular momentum operator commutes with the atomic and molecular Hamiltonian. It is quantized in units of $J(J+1)$, so that

$$\epsilon(\text{rot}) = BJ(J+1) \tag{5.21}$$

where J is the total angular momentum quantum number and it is equal to 0, 1, 2, 3, etc.

When such a rotating molecule is placed in a magnetic field, one can detect a molecular magnetic moment that is generated by the rotation. Both the nuclei and electrons are charged and their rotation should generate magnetic moments of the opposite sign. The magnetic moment due to the nuclei is very easy to calculate from Eq. (1.3) since the angular rotation of the nuclei is

$$\boldsymbol{\omega}(\text{rot}) = \frac{h\mathbf{J}}{2\pi I}, \tag{5.22}$$

so that

$$\boldsymbol{\mu}(\text{nuclei}) = \frac{eh\mathbf{J}}{4\pi cI}\sum_i Z_i z_i^2 \tag{5.23}$$

where Z_i is the charge number of each nucleus that is located z_i from the center of mass of the molecule. Since $\boldsymbol{\mu}$ is proportional to **J** we can define a g value due to the rotating nuclei, and if

$$\boldsymbol{\mu}(\text{nuclei}) = g(\text{nuclei})\mu_N\mathbf{J}, \tag{5.24a}$$

then

$$g(\text{nuclei}) = \frac{m_p}{I}\sum_i Z_i z_i^2 \tag{5.24b}$$

where m_p is the proton mass. The nuclear magneton μ_N is used in Eq. (5.24*a*) so that these g values will be reasonably close to unity. In fact, for H_2, a glance at Eq. (5.24*b*) shows that $g(\text{nuclei}) = 1.0000$ for the rotational magnetic moment of its two nuclei.

CONTRIBUTION OF ELECTRONS

The contribution of the electrons to the rotational magnetic moment is much more difficult to establish, although the final formula is relatively simple. In his early papers G. C. Wick (1933) showed how an inverted form of the Larmor relation could give the correct answer. The problem is that the electrons do not form a rigid cloud as do the nuclei and one must account for the combined effects of the rotation and the magnetic field. Although Wick's reasoning does this, we prefer the more fundamental approach of J. R. Eshbach and M. W. P. Strandberg (1952).

For a rotating molecule Eq. (1.18) must be modified so that

$$\mathsf{H} = \mathsf{H}_0 + \mathsf{H}(\mathrm{rot}) + \mathsf{H}'(H). \tag{5.25}$$

Since the rotational energies are much greater than the magnetic ones, one must first consider the effects of $\mathsf{H}(\mathrm{rot})$ upon H_0 before the perturbation by $\mathsf{H}'(H)$. Since the total angular momentum $\mathbf{J}$ has a small contribution due to the electronic orbital angular momentum, the effect of $\mathsf{H}(\mathrm{rot})$ is quite important. Without rotation the orbital angular momentum was quenched, so that matrix elements of the type $\langle\psi_i^0|\, \mathsf{L}_Z\, |\psi_i^0\rangle$ are zero. With the addition of $\mathsf{H}(\mathrm{rot})$ the electrons must be measured in the molecular x, y, and z coordinates, which are rotating with the nuclei. The combined effects of H_0 and $\mathsf{H}(\mathrm{rot})$ are that the wave functions ψ_i^0 have a first-order dependence upon $\mathbf{J}$. We must use ψ_i^J to evaluate the matrix elements for $\mathbf{L}$ and

$$\langle\psi_0^J|\, \mathsf{L}_x\, |\psi_0^J\rangle = 4B\mathsf{J}_x \sum_{n\neq 0} \frac{|\langle\psi_0^0|\, \mathsf{L}_x\, |\psi_n^0\rangle|^2}{\epsilon_n^0 - \epsilon_0^0} \tag{5.26}$$

where x will be assumed to be perpendicular to the axis of the linear molecule. The form of Eq. (5.26) is such that the rotation gives a rotation-induced magnetic moment which is very similar in form to the field-induced term $\epsilon^{(2)}H$. Both of these induced moments arise from a second-order perturbation of H_0 in Eq. (5.25), but the rotation-induced moment gives an energy that is both linear in the rotation and linear in the field. An orbital moment is defined by

$$\boldsymbol{\mu}(\text{electrons}) = -\mu_B\mathbf{L} \tag{5.27a}$$

and it corresponds to $\mu_{\|}$ in the vector model, so that

$$\boldsymbol{\mu}(\text{electrons}) = g(\text{electrons})\mu_N\mathbf{J}. \tag{5.27b}$$

Then

$$g(\text{electrons}) = -\frac{4Bm_p}{m} \sum_{n\neq 0} \frac{|\langle\psi_0^0|\, \mathsf{L}_x\, |\psi_n^0\rangle|^2}{\epsilon_n^0 - \epsilon_0^0} \tag{5.28}$$

where $m_p/m = \mu_B/\mu_N$.

In determining the electronic contribution Eq. (5.28) must account for at least two factors. Some of the electrons in a molecule will follow the rotation exactly and they will cancel part of $\boldsymbol{\mu}$(nuclei). At the same time, however, some of the electrons will tend to "slip" as the nuclei rotate and they will give a decreased cancellation. In fact, it is possible for the electronic contribution to slightly exceed the nuclear one since the electrons that follow can have a larger radius of rotation than the nuclei. It would be expected in H_2 that, since its two electrons are engaged in bonding, they would tend to "slip" as the molecule rotates, so that its total g(rot) value should be moderately close to the value for the nuclei. Table 5.3 lists some g(rot) values determined for simple molecules.

Table 5.3. *Rotational Magnetic Moments for Simple Molecules*

Molecule	g(rot)[a]
H_2	+0.88291
DH	+0.66321
D_2	+0.44288
^{7}LiH	−0.642
$H^{19}F$	+0.7392
$D^{19}F$	+0.3695
$^{12}C^{16}O$	−0.2691
$^{14}N^{16}O(^{2}\Pi_{1/2})$	−42.0[b]
$^{16}O^{12}C^{32}S$	−0.025
$^{12}CH_4$	+0.3133[c]

[a] Sum of g(nuclei) and g(electrons).

[b] For NO in its $^{2}\Pi_{1/2}$ ground state one has $\mu_{\parallel} = 0$ without rotation [see (Eq. 4.11*a*)]. With rotation Eq. (5.26) gives a mixing of the $^{2}\Pi_{1/2}$ and $^{2}\Pi_{3/2}$, so that $\mu_{\parallel}$ is no longer zero and contributes to g(rot).

[c] Although CH_4 is tetrahedral and not linear, it has sufficiently high symmetry to have only one g(rot) value.

One can see from Table 5.3 that most of the electrons in large molecules must tend to follow the rotation. This is not surprising since most of the electrons in molecules occupy tightly bound orbitals around a single nucleus. However, the "slip" of some of the electrons is still important, particularly in the bonding electrons, as is clearly illustrated in H_2. The very large electron contribution for NO is a special case. This is because rotation yields a rapid transition from Hund's case A to case B in this $^2\Pi$ molecule (see Chapter 4).

RELATION TO χ_M

If the structure of a linear molecule is known, g(nuclei) can be easily calculated. And if g(rot) is experimentally determined, we can evaluate g(electrons) from the difference. Comparison of Eqs. (5.28) and (5.18) shows that to evaluate the paramagnetic part of χ_M from g(electrons) we must determine the matrix elements for L_Z averaged over all molecular orientations. The matrix elements in Eq. (5.28) are for the molecule-fixed coordinates, and if we consider the orientations of these coordinates in the space-fixed Z axis,

$$\begin{aligned} |\langle 0|\, \mathsf{L}_Z \,|n\rangle|^2_{\text{av}} &= \tfrac{1}{3}\, |\langle 0|\, \mathbf{L} \,|n\rangle|^2 \\ &= \tfrac{1}{3}\{|\langle 0|\, \mathsf{L}_x \,|n\rangle|^2 + |\langle 0|\, \mathsf{L}_y \,|n\rangle|^2 + |\langle 0|\, \mathsf{L}_z \,|n\rangle|^2\}. \end{aligned} \tag{5.29}$$

In a $^1\Sigma$ linear molecule with z as the linear axis, reasoning similar to Eq. (5.1) shows that

$$|\langle 0|\, \mathsf{L}_z \,|n\rangle|^2 = 0 \tag{5.30a}$$

and

$$|\langle 0|\, \mathsf{L}_x \,|n\rangle|^2 = |\langle 0|\, \mathsf{L}_y \,|n\rangle|^2. \tag{5.30b}$$

Combining Eqs. (5.28), (5.29), (5.30), and (5.18) we can write

$$\chi_M = -\frac{Ne^2}{6mc^2}\left[\sum_i \langle r^2(i)\rangle_{\text{av}} + \frac{Ig(\text{electrons})}{m_p}\right] \tag{5.31}$$

where $r(i)$ must be measured with respect to the center of mass. This relation is only valid as written for linear molecules, although with a slight modification it can be also applied to nonlinear ones.

This relation between g(electrons) and the paramagnetic part of χ_M is very convenient since it is the paramagnetic part that is so difficult to calculate. The values of g(rot) can either be determined from spectroscopic studies on molecular beams or from more conventional spectroscopy with the application of strong magnetic fields. The most interesting thing about Eq. (5.31) is that it shows the interrelation of two fundamental magnetic properties of molecules. One of the most important tasks in physical chemistry is to find such relations and to show that many physical properties which seem to be unrelated are actually dependent one upon the other.

5.4. Ramsey's Equation for Nuclear Shielding

The shielding of a nucleus is much more complex in a molecule than it is for an atom. In a diamagnetic atom there is only a diamagnetic moment, and the Lamb equation illustrates that the resultant shielding can be easily calculated. N. F. Ramsey (1950) showed how the shielding in molecules depends upon the paramagnetic moment, but the resultant expression is rather complicated. However, as a result of nuclear magnetic resonance spectroscopy, it is possible to measure the shielding or "chemical shift" of nuclei in a wide variety of molecules, and the theoretical treatment of this subject is one of considerable interest.

When a molecule is oriented in a magnetic field such that the field is along a principal axis of the susceptibility tensor, Eq. (5.15) for an atom can then be generalized so that

$$H = H_0(1 - \sigma_g) \tag{5.32}$$

where $g = x$, y, or z depending upon the molecular orientation. The average value of σ, appropriate to molecules in solution, would be given by

$$\sigma = \tfrac{1}{3}(\sigma_x + \sigma_y + \sigma_z). \tag{5.33}$$

The internal field due to the induced moment in a molecule can be obtained from Eq. (5.14), and for H_0 along the molecular z axis

$$\begin{aligned} H(\text{loop}) &= -\sigma_z H_0 \\ &= -\frac{e^2 H}{2mc^2}\left\langle\frac{x^2 + y^2}{r^3}\right\rangle - 2\mu_B\left\langle\frac{\mathsf{L}_z}{r^3}\right\rangle. \end{aligned} \tag{5.34}$$

The first term in H(loop) is derived from the diamagnetic or Larmor moment and the second from the paramagnetic one. As we saw in the Lamb equation, the Larmor term can be determined with this expression and with the unperturbed wave function of the molecule. The paramagnetic term, on the other hand, is zero unless we use perturbed wave functions. The perturbation that is responsible for a nonzero value for $\langle \mathsf{L}_z \rangle$ is the term $\epsilon^{(2)}H^2$ in the energy. This term results from the mixing of the ground-state wave function with those in excited electronic states by the magnetic field. This mixing gives Ramsey's equation, and if we sum over all the electrons and set $H = H_0$ whenever appropriate,

$$\sigma_z = \frac{e^2}{2mc^2} \sum_i \left\langle \frac{x^2(i) + y^2(i)}{r^3(i)} \right\rangle - 2\mu_B^2 \sum_i \sum_{n \neq 0} \left\{ \frac{\langle 0 | \mathsf{L}_z(i) | n \rangle \langle n | [\mathsf{L}_z(i)/r^3(i)] | 0 \rangle}{\epsilon_n^0 - \epsilon_0^0} + \frac{\langle 0 | [\mathsf{L}_z(i)/r^3(i)] | n \rangle \langle n | \mathsf{L}_z(i) | 0 \rangle}{\epsilon_n^0 - \epsilon_0^0} \right\}. \tag{5.35}$$

The Larmor term is slightly more complicated than in an atom only because molecules are not spherically symmetric. The paramagnetic term, however, is quite complicated. The matrix elements $\langle 0 | \mathsf{L}_z | n \rangle$ enter into the expression because they are the ones that mix the ground state with excited states. A necessary condition so that the paramagnetic term will not be zero is that neither $\langle 0 | \mathsf{L}_z | n \rangle$ nor $\langle n | \mathsf{L}_z/r^3 | 0 \rangle$ be equal to zero and this is clearly expressed in Ramsey's result. Similar expressions can be written for σ_x and σ_y, and to obtain the average σ all three values must be evaluated. Since the induced paramagnetic moment is aligned with the field, the paramagnetic term in Eq. (5.35) tends to oppose that due to diamagnetism. This, of course, is not a surprising result.

The experimental value for the shielding of the protons in H_2 is $\sigma = 26.4 \pm 0.6$ ppm. The diamagnetic shielding can be rather accurately calculated and one obtains $\sigma(\text{dia}) = 32.1$ ppm. This means that $\sigma(\text{para}) \approx -5.7$ ppm. One can see that for H_2 the paramagnetic contribution to shielding is small, but it cannot be neglected. It has also been found that, like the paramagnetic contribution to χ_M, the paramagnetic contribution to σ can be related to an experimental spectroscopic parameter. The parameter in this case is called a spin-rotation constant c_i. The spin-rotation constant determines the amount of nuclear splitting that is produced by the rotation of a molecule. It is essentially a measure of the magnetic field, due to rotation, which is produced at the nucleus. We shall not elaborate further on spin-rotation interaction other than to say that $\sigma(\text{para})$ can be related to c_i in a way very closely related to Eq. (5.31). With this relationship and the accurately determined spin-rotation constant c_H in H_2 one can show that $\sigma(\text{para}) = -5.5$ ppm, in excellent agreement with experiment. In a similar comparison for the protons in CH_4 the work cited in Table 5.1 has shown that $\sigma(\text{para}) = -57.1$ ppm from spin-rotation measurements. Their theoretical value with the Pitzer wave function was $\sigma(\text{dia}) = 88.0$ ppm, and so $\sigma(\text{theory}) = 30.9$ ppm. The experimental value for the shielding of the protons in CH_4 is $\sigma = 30.6 \pm 0.6$ ppm.

CHEMICAL SHIFT

In nuclear magnetic resonance work one usually measures the relative change in the shielding when a single nucleus is examined in a series of molecules. This is called the chemical shift. The value of the chemical shift depends not so much upon the absolute value of σ but upon how it changes between molecules.

The dependence upon the radius in Eq. (5.35) shows that the electrons close to the nucleus would be expected to be the most important ones in determining the chemical shift. This is probably true for the most common nuclei except for hydrogen. The hydrogen nucleus is essentially surrounded by only one electron and this electron is close, in most molecules, to being in a $1s$ orbital. In molecules such as CH_4, and CH_3Cl the large number of electrons associated with the other nuclei must contribute to σ for the protons. For example, the Lamb equation predicts a $\sigma = 18$ ppm for a $1s$ electron around a proton but this is only 20 per cent of $\sigma(\text{dia})$ calculated for the protons in CH_4.

In the case of a nucleus like fluorine, the dominant contribution to σ comes from its own electrons. In addition, in most molecules, the dominant contribution to the chemical shift comes from changes in the paramagnetic contribution. As an example, F^- is a closed-shell S state and the paramagnetic contribution to σ for it is exactly zero. On the other hand, F_2 has a structure close to the fluorine atom and one would expect a large paramagnetic contribution for F_2. It is found by experiment that the σ for F_2 is over 500 ppm more paramagnetic than is aqueous F^-. Figure 5.4 shows the correlation between the increase in the diamagnetism of the shielding and the electronegativity difference between fluorine and a single-bonding atom. Compounds with greater ionic character are found to have what appears to be a smaller paramagnetic contribution to the shielding. This is in excellent agreement with the simple reasoning that ionic molecules should approach F^- and have no paramagnetic contribution to σ.

When chemical shifts for fluorine and other nuclei are examined in close detail, such simple reasoning usually fails. The study of chemical shifts is primarily based upon empirical reasoning and not upon quantitative use of the Ramsey equation. One of the best qualitative models is to consider the effect of the electrons on neighboring atoms in terms of the magnetic field produced by induced magnetic dipole moments. In Fig. 5.3 a field produced at the protons in benzene is shown which is due to

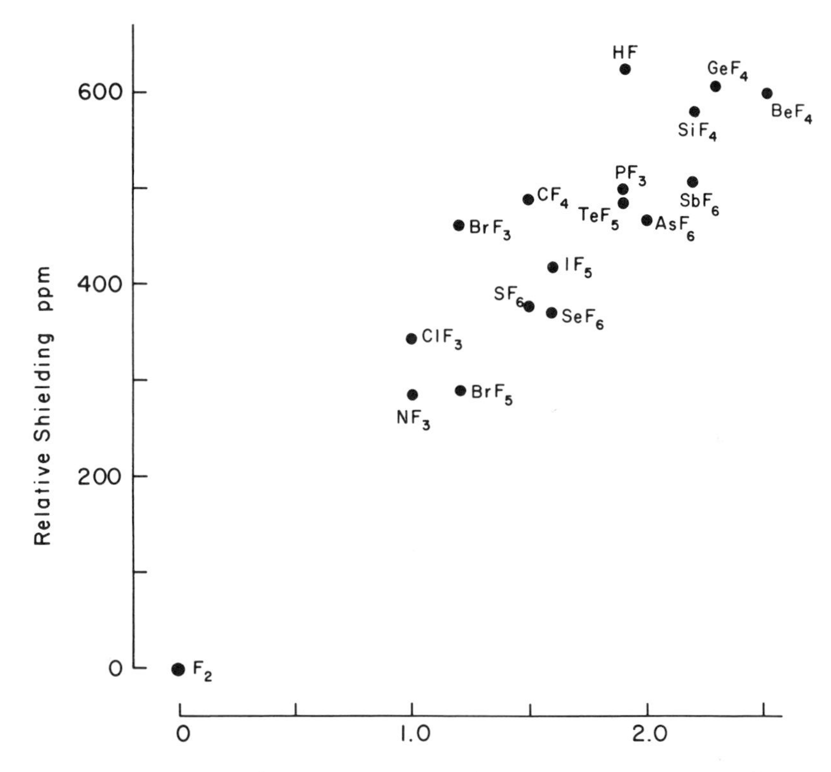

Figure 5.4. *Increase in diamagnetic shielding for fluorine nuclei with the electronegativity difference in simple molecules. [After H. S. Gutowsky and C. J. Hoffman, J. Chem. Phys.,* **19,** *1259 (1951).]*

the field-induced moment of the p_z carbon electrons. When the magnetic field is along the molecular z direction, this field is in the direction of the applied field. If the applied field were along the plane of the benzene ring, there would be no large ring-current field at the protons. The net effect for a benzene molecule tumbling in a solution would be to give an average paramagnetic contribution to the shielding of the protons. It is observed in NMR that the benzene protons have such a chemical shift relative to similar molecules that are not aromatic.

If it is assumed that the effect of neighboring atoms can be accounted for in terms of induced dipole moments, then Fig. 5.5 shows the field produced by such dipole moments. The equations for the dipole field assume that r is much larger than the dimensions of the dipole moment. Since an induced dipole moment is at least as large as an atom, the dipole

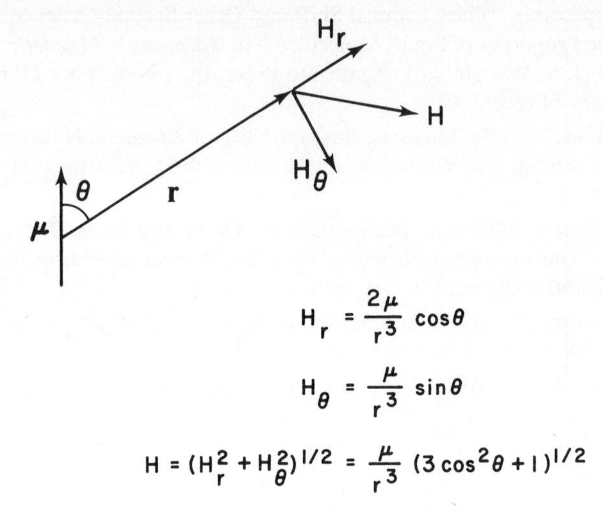

$$H_r = \frac{2\mu}{r^3}\cos\theta$$

$$H_\theta = \frac{\mu}{r^3}\sin\theta$$

$$H = (H_r^2 + H_\theta^2)^{1/2} = \frac{\mu}{r^3}(3\cos^2\theta + 1)^{1/2}$$

Figure 5.5. *Field due to a dipole moment. This assumes that r is much greater than the dimensions of the dipole.*

field can be only a rough approximation for neighboring atoms in a molecule. However, we can see that if the magnitude of the induced moment is independent of the molecular orientation, then, as the molecule tumbles, the dipole field on a neighboring atom will average to zero since $\cos\theta$ and $\sin\theta$ average to zero. An anisotropic induced moment will not average to zero, and benzene is an example of a highly anisotropic induced moment. In acetylene, where the protons lie along the axis of an anisotropic induced moment, one would predict a large diamagnetic shift and this is also observed.

The neighboring atoms can have induced moments that are anisotropic in a number of different types of molecules, but care must be taken in using this kind of argument since at best it only accounts for the effect of neighboring atoms. In tetramethylsilane (TMS) one finds that its protons have a relatively large diamagnetic shielding. One is tempted to ascribe this to the electron-rich silicon atom. However, since the silicon is in a site of essentially tetrahedral symmetry its own magnetic moment must be isotropic, but the induced moments on the carbon atoms can, and probably do, contribute to the shielding in TMS.

References

D. W. Davies, *The Theory of the Electric and Magnetic Properties of Molecules*, John Wiley & Sons, Inc., New York (1957). A modernized Van Vleck, particularly strong in diamagnetism. Written for graduate students.

W. N. Lipscomb, "The Chemical Shift and Other Second-Order Magnetic and Electric Properties of Small Molecules," in *Advances in Magnetic Resonance*, Vol. 2 (J. S. Waugh, ed.), Academic Press, Inc., New York (1966). A nice summary of recent work.

J. I. Musher, "On the Magnetic Susceptibility of Aromatic Hydrocarbons and 'Ring Currents,'" *J. Chem. Phys.*, **43,** 4081 (1965). An attack on the concept of ring currents.

N. F. Ramsey, *Molecular Beams*, Oxford University Press, Inc., New York (1963). The authoritative work; includes nuclear shielding, rotational g values, and spin-rotation interaction.

N. F. Ramsey, "Nuclear Interactions in Molecules," *Amer. Scientist*, **49,** 509 (1961). A short popular summary.

L. Salem, *The Molecular Orbital Theory of Conjugated Systems*, W. A. Benjamin, Inc., Menlo Park, Calif. (1966). Contains a clear exposition of the molecular-orbital theory of ring currents.

Problems

5.1. In a Bohr atom the electrons rotate, so that their angular momentum is a multiple of $\hbar$. Assume an electron radius equal to a_0 and an angular momentum equal to $\hbar$. Compare the resulting angular velocity to that of the Larmor precession when $H = 10^4$ gauss.

5.2. Show that Van Vleck's gauge can be defined by the equation $\mathbf{A} = \frac{1}{2}\mathbf{H} \times \mathbf{r}$. Also show that if this radius r is defined with an arbitrary coordinate center this still satisfies the condition that $\mathbf{H} = \text{curl } \mathbf{A}$. [*Note:* This coordinate transformation is a commonly used gauge transformation in molecules. One can show that χ_M(dia) is a maximum when $\mathbf{r}$ is located at the center of electronic charge.][See S. I. Chan and T. P. Das, *J. Chem. Phys.*, **37,** 1527 (1962).]

5.3. Show that the gauge $A_X = 0$, $A_Y = XH$ and $A_Z = 0$ satisfies the basic equation $\mathbf{H} = \text{curl } \mathbf{A}$ where H is along Z. Try to formulate $\mathsf{H}'(H)$ for this choice of gauge using the same method as Section A.4. Since the total energy contributed by $\mathsf{H}'(H)$ must be independent of the choice of gauge, show that both the "diamagnetic" and "paramagnetic" terms must contribute for atoms with this choice of gauge. This clearly shows that gauge has a great deal to do with the use of the terms "diamagnetic" and "paramagnetic" in theoretical equations.

5.4. Use the observed value for χ_M for He to calculate the mean-square radius for each of its electrons in angstrom units. Compare this value with that calculated from hydrogenlike wave functions where for ns wave functions $\langle r^2 \rangle = [n^2 a_0^2(5n^2 + 1)]/2Z^2$. Electron–electron repulsion is, of course, important in He. Use the observed value of $\langle r^2 \rangle$ to determine an effective Z value and establish the screening of the nuclear charge by the second electron by determining the best Z value.

5.5.* The hydrogenlike-wave-function expression is $\langle 1/r \rangle = Z/a_0n^2$. Use this expression and the Lamb equation to calculate σ in ppm for the He atom. Either use the shielding found in Problem 5.4 or use the variation value of $Z = \frac{27}{16}$. (*Note:* The accurate self-consistent-field value is $\sigma = 60$ ppm.)

5.6. Although CH_4 has a neonlike structure, it should have a much larger value for $\langle r^2 \rangle$ because of the effect of the four protons. This is confirmed by comparing $\chi_M = -6.74 \times 10^{-6}$ as observed for Ne compared with χ_M(dia) calculated in Table 5.1 for CH_4. Since the two "inner" orbitals in CH_4 are shielded from the proton, one might expect that the $\langle r^2 \rangle$ for the $1a_1$ and $2a_1$ in CH_4 to be similar to that expected for $1s$ and $2s$ in an atom. Use the observed susceptibility for Ne to calculate $\langle r^2 \rangle$ for all its electrons. Compare this value with that calculated for CH_4 and show that the Ne value is close to that calculated for only the two $2a_1$ electrons in CH_4. This is reasonable, for in an atom the $\langle r^2 \rangle$ for a $2p$ electron is smaller than it is for a $2s$.

5.7.* The observed anisotropic values for single crystals of benzene at 261°K are $\chi_M^a = -65.2$, $\chi_M^b = -37.9$, and $\chi_M^c = -61.3 \times 10^{-6}$. Use these values to calculate the average or isotropic χ_M to be observed in polycrystalline samples. Compare this with the χ_M in Table 1.1 given for liquid near 293°K. (*Note:* Since benzene forms orthorhombic crystals, the a, b, and c axes are at right angles.) For a general crystal system the principal values of χ_M are designated χ_M^1, χ_M^2, and χ_M^3.

5.8. The crystal structure of benzene is illustrated in Fig. 5.2. The angles between the perpendiculars to the planes of the molecules and the a, b, and c axes are 44°49′, 77°04′, and 48°04′, respectively. The crystal contains four different kinds of molecules, but in this simple orthorhombic crystal, although all four of the benzene molecules have different orientations, they form the same angles with the crystal axes. In this case, the values for χ_M^a, χ_M^b, and χ_M^c given in Problem 5.7 can all be related to the χ_M^x, χ_M^y, and χ_M^z of a single molecule. Because of the symmetry in benzene $\chi_M^x = \chi_M^y$. The relations between the crystal and molecular susceptibilities are

$$\chi_M^a = \chi_M^x + (\chi_M^z - \chi_M^x)\cos^2\theta_{az},$$

$$\chi_M^b = \chi_M^x + (\chi_M^z - \chi_M^x)\cos^2\theta_{bz},$$

and

$$\chi_M^c = \chi_M^x + (\chi_M^z - \chi_M^x)\cos^2\theta_{cz}$$

where θ_{az}, θ_{bz}, and θ_{cz} are the angles between the z and a, b, and c axes, respectively. Use these relations to solve for χ_M^x and χ_M^z. Compare your results to those given in Table 5.2. The values calculated in this problem are probably more accurate because of the better crystal-structure data given here.

5.9. Use the values determined in Problem 5.8 to calculate $\langle \rho^2 \rangle^{1/2}$ for benzene. For $\chi_M(\pi)$ either use $\chi_M^z - \chi_M^x$ or correct χ_M^z for the other electrons, using Pascal's χ_A values.

5.10. Derive the relations between χ_M^a, χ_M^b, χ_M^c and χ_M^x, χ_M^y, χ_M^z given in Problem 5.8. The best way to do this is to form the direction-cosine matrix to transform a vector from the x, y, and z system into the a, b, and c one. The susceptibility matrix can be transformed from one axis system into the other by this transformation together with its inverse. In this way the direction-cosine matrix forms a unitary transformation. Do not forget the orthogonality and normality relations of a unitary transformation.

5.11.* Show that while H_2 and D_2 have different moments of inertia and different values for g(rot) they should have very nearly the same value for χ_M in Eq. (5.31). Can you also show that HD has the same value for χ_M from this equation?

5.12. Calculate g(nuclei) for $H^{19}F$ and $^{12}C^{16}O$. The internuclear distances are H—F (0.92 Å) and C—O (1.13 Å). For HF, estimate g(rot) by assuming that it has an ionic bond and is H^+F^-. Compare your estimate with the value given in Table 5.3.

5.13. Calculate g(nuclei) for $^{16}O^{12}C^{32}S$. This is a linear molecule that has bonding close to that of CO_2. From rotational spectroscopy it is found that the internuclear distances are O—C (1.16 Å) and C—S (1.56 Å). Since $\chi_M = -32.4 \times 10^{-6}$, compare the diamagnetic and paramagnetic parts to χ_M, using the value for g(rot) given in Table 5.3.

5.14. A tetrahedral molecule like CH_4 has only a single moment of inertia and a single value for g(rot). It can be treated like a linear molecule except that all three terms on the right side of Eq. (5.29) are equal. Modify Eq. (5.31) for CH_4, show that g(nuclei) $= 1.0000$, and confirm the value for χ_M(para) assumed in Table 5.1.

5.15. Show that if r is constant in Eq. (5.35) and can be factored out of all the matrix elements, the result is consistent with the equation $\sigma = -2\mu(\text{induced})/r^3$. The first term in Eq. (5.35) gives the diamagnetic induced moment and the second the paramagnetic one. Do not forget that the paramagnetic one was calculated in Chapter 1 and can be found using Eq. (1.22).

Solutions:

5.5. If we take $Z = \frac{27}{16}$ and $n = 1$, we obtain from Eq. (5.17) for two electrons

$$\sigma = \frac{54e^2}{48mc^2a_0}$$

$$= \frac{54 \times (4.8 \times 10^{-10})^2}{48 \times 9.1 \times 10^{-28} \times 9 \times 10^{20} \times 0.505 \times 10^{-8}}$$

$$= 63 \times 10^{-6}.$$

5.7. For a polycrystalline sample

$$\chi_M = \left(\frac{\chi_M^X + \chi_M^Y + \chi_M^Z}{3}\right).$$

Since benzene forms orthorhombic crystals, we can associate X, Y, and Z with a, b, and c. This gives

$$10^6\chi_M = \left(\frac{-65.2 - 37.9 - 61.3}{3}\right)$$
$$= -54.6.$$

The liquid value is -54.85.

5.11. For a diatomic molecule composed of m_1 and m_2 which are z apart,

$$I = \left(\frac{m_1 m_2}{m_1 + m_2}\right) z^2.$$

Since H_2 and D_2 both have the center of mass halfway between the nuclei, the first term in Eq. (5.31) should be the same for the two molecules. The moment of inertia

$$\frac{I(D_2)}{I(H_2)} = \frac{m(D)}{m(H)} = 1.9985$$

and from Table 5.3

$$\frac{g(H_2)}{g(D_2)} = \frac{0.88291}{0.44288} = 1.9936.$$

As a result, Eq. (5.31) should have both terms almost identical for H_2 and D_2.

For HD the center of mass is not halfway between the nuclei and its value for $\langle r^2 \rangle$ is not measured in the same coordinate system as for H_2 and D_2. Except for rotational centrifugal distortion and vibrational averaging we expect that χ_M should be the same for all isotropic species.

6

Fundamentals of Magnetic Resonance Spectroscopy

THE MAGNETIC MOMENTS that yield paramagnetism in bulk samples can also be studied by electron paramagnetic resonance (EPR) spectroscopy. This method measures the energy-level splittings that are produced when a system of permanent molecular magnetic moments is placed in a magnetic field. These same splittings can also be related to the paramagnetic susceptibility of these systems, and EPR is a tool that both complements and supplements the work discussed in Chapters 1 through 4.

Many nuclei possess magnetic moments, and it is also possible to measure the splittings they produce in a magnetic field. This form of spectroscopy is called nuclear magnetic resonance (NMR). Since the nuclear moments are small, it is nearly impossible to determine them from magnetic susceptibility measurements, but NMR is an important tool for the investigation of diamagnetic molecules.

The use of three-letter abbreviations for these two forms of spectroscopy has produced a certain amount of confusion. The importance of the spin magnetic moment has led to electron spin resonance (ESR) in place of EPR. Since much NMR is done on protons, one finds PMR used for proton magnetic resonance. The latter can be confused with paramagnetic magnetic resonance. Electron magnetic resonance (EMR) is a logical term but it is rarely used.

Both NMR and EPR have many features in common. The experimental apparatus has some common aspects and, above all, their theories are strongly related. This chapter is devoted to their interrelations and to their fundamental differences.

6.1. Zeeman Energy Levels

The energy of a magnetic moment in a magnetic field is expressed by the Zeeman part of the Hamiltonian and

$$\mathsf{H}(\text{Zeeman}) = -\mu \cdot \mathbf{H}. \tag{6.1}$$

The nuclear moment is defined in Eq. (1.9*c*), so with H along the Z axis we can write

$$\epsilon(\text{nuclear-Zeeman}) = -\frac{\mu_I}{I}\mu_N H M_I \tag{6.2}$$

where $M_I = -I, \; -I+1, \ldots, I$. This equation assumes that I_Z is exactly quantized (or that I_Z commutes with the molecular Hamiltonian) and this is often, but not always, true.

Since most molecules have quenched orbital angular momentum, we can follow the spin-Hamiltonian formalism and according to Eq. (3.28*c*)

$$\epsilon(\text{electron-Zeeman}) = g\mu_B H M_S \tag{6.3}$$

where g may be close to 2.0; it always contains, however, the contribution of the quenched orbital moment. Again we have had to assume that S_Z can be taken as exactly quantized, and, for electrons, we have neglected the energy shift that results from second-order paramagnetism since it is independent of M_S. If we use γ, the magnetogyric ratio, one equation can be written for both kinds of spin, and with Eq. (1.4)

$$\epsilon(\text{Zeeman}) = -\frac{\gamma h H}{2\pi} m \tag{6.4a}$$

where $m = M_S$ or M_I with

$$\gamma(\text{electron spin}) = -\frac{2\pi g\mu_B}{h} = -\frac{ge}{2m_e c} \tag{6.4b}$$

$$\gamma(\text{nuclear spin}) = \frac{2\pi g_I \mu_N}{h} = \frac{2\pi\mu_I\mu_N}{Ih}. \tag{6.4c}$$

The energy levels predicted by Eq. (6.4*a*) consist of a single ladder with two levels for spin = ½ and three equally spaced for spin = 1, etc., as

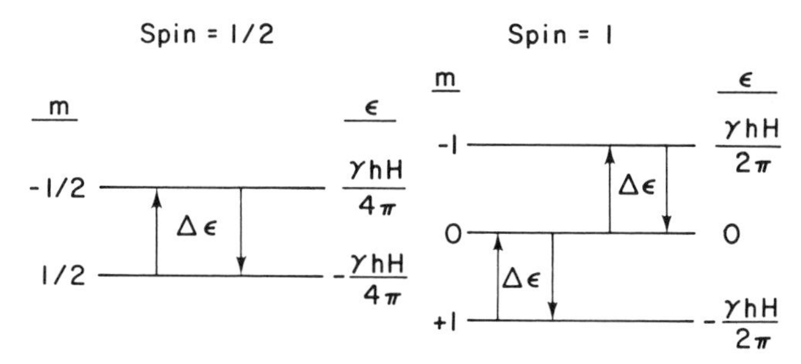

Figure 6.1. *Energy levels for two possible spin values and transitions which follow* $\Delta m = \pm 1$ *selection rule. For electrons, since* γ *is negative, the entire pattern is inverted.*

shown in Fig. 6.1. If we consider the spin = ½ case, there is only one possible spectroscopic transition, with

$$h\nu = \Delta\epsilon = \frac{\gamma h H}{2\pi}$$

and

$$\nu = \frac{\gamma H}{2\pi} \tag{6.5a}$$

or

$$\omega = 2\pi\nu = \gamma H. \tag{6.5b}$$

One can see that this expression is also true for spin = 1, etc., if transitions are only possible between adjacent energy levels. These frequencies for nuclei are listed in Section A.3 for fields of 10,000 gauss.

Although Eq. (6.5) may express the frequency of a magnetic resonance transition, it contains no information about the width or complexity of the transition. In some branches of spectroscopy one is satisfied to simply observe spectroscopic transitions and little work is done on the width of the observed spectroscopic lines. However, in magnetic resonance spectroscopy the width or line shape of the observed transitions is very important and much work in magnetic resonance is entirely concerned with this aspect of the field. The quantum-mechanical theory of spectroscopic transitions is one of the more difficult fields in quantum mechanics and we shall not touch upon this. Fortunately there is an easy-to-understand classical theory of magnetic resonance and this theory is the starting point of most serious discussions in the field. The bulk of this chapter will be concerned with this important theory.

6.2. Free Precession of Spins

In addition to magnetic moments, electrons and nuclei also possess angular momenta. In fact, γ is the ratio between these two simultaneous properties. In a magnetic field a magnetic moment experiences a force that tries to align the moment. However, one of the properties of an angular momentum is that it reacts to an applied force to yield a resultant that is at right angles to the applied force. This problem is essentially that found for the effect of a gravitational force on a gyroscope. The result is that the gyroscope precesses about the gravitational field. In the same way the angular momenta of electrons and nuclei precess about a magnetic field.

The classical equation of motion for a magnetic moment and its associated angular momentum in a magnetic field is given by

$$\frac{d\boldsymbol{\mu}}{dt} = \gamma\boldsymbol{\mu} \times \mathbf{H}. \tag{6.6}$$

If $\mathbf{H}$ is along the Z axis, then in terms of components there are three equations,

$$\frac{d\mu_X}{dt} = \gamma\mu_Y H, \tag{6.6a}$$

$$\frac{d\mu_Y}{dt} = -\gamma\mu_X H, \tag{6.6b}$$

and

$$\frac{d\mu_Z}{dt} = 0. \tag{6.6c}$$

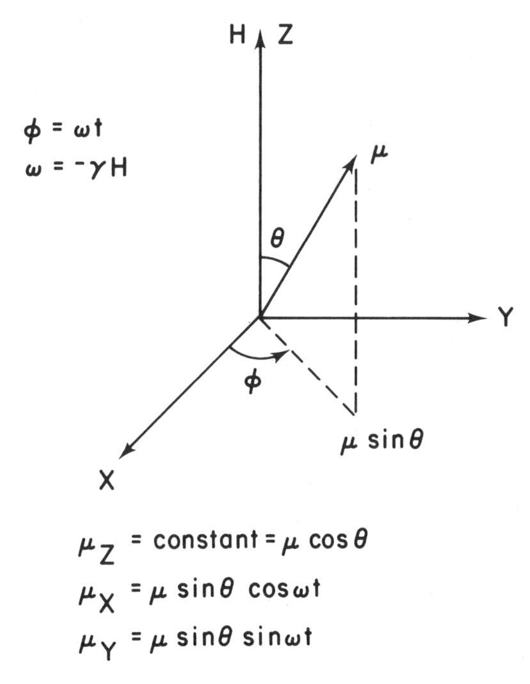

Figure 6.2. *Free procession of a magnetic moment in a magnetic field.*

The solution to these equations is shown in Fig. 6.2 so that for the angular velocity of precession

$$\omega = \frac{d\phi}{dt} = -\gamma H. \tag{6.7}$$

The result for the angular precession of the magnetic moment of an electron given by Eq. (6.7) is very similar to that for the precession of the charge of the electrons in an atom given by Eq. (5.7*b*). They are exactly related; since $\gamma(\text{orbital}) = -e/2mc$, Eq. (6.7) is then a general equation for both precessional frequencies. Both frequencies are commonly called the Larmor frequency, although some confusion results from calling both kinds of precession the Larmor precession. It is clear from Fig. 6.2 that we are here concerned with a precession of the magnetic moment of the electron. The Larmor precession of Chapter 5 referred to an apparent precession of the charge of the electrons in an atom.

One of the most remarkable things about the frequency given by Eq. (6.7) is that it is identical with the spectroscopic frequency given by Eq. (6.5). It is the identity between these two frequencies which justifies the use of resonance in the designation magnetic resonance. All of spectroscopy is involved with a "resonance" between electromagnetic radiation and energy-level spacings, but in magnetic resonance there is a further "resonance" between the radiation and the precessional motion of the spins. The identity of these two frequencies suggests that there should be a simple picture, based upon classical mechanics, for the spectroscopy of spin systems. This simple picture should also be consistent with quantum mechanics as summarized in the vector model.

The Larmor frequency given by Eq. (6.7), which is a result of classical mechanics, is completely consistent with quantum mechanics. The vector-model diagram of Fig. 2.1 is that of a precession of the angular momentum about the Z axis, and one can show from quantum mechanics that for a magnetic moment in a magnetic field its precession is quantized and given by Eq. (6.7). This does not mean that the precessional frequency in the vector model is always given by Eq. (6.7), for this is the correct result only for systems where the Hamiltonian is of the form of Eq. (6.1).

For a collection of spins only the properties of the total magnetic moment or magnetization **M** can be measured, particularly if there is any interaction at all between the spins. For the total magnetic moment we can again use Eq. (6.6), since M is just the vector sum of the individual magnetic moments all with the same γ value, so that

$$\frac{d\mathbf{M}}{dt} = \gamma\mathbf{M} \times \mathbf{H}. \tag{6.8}$$

The possible solutions to this equation are much more complicated than was Eq. (6.6). If we assume that somehow a group of individual magnetic moments are precessing in phase, they will produce an **M** that will also precess, and this **M** will follow Eq. (6.6). The precession of this **M** will continue until something happens to break up or dephase the spins. For a group of random spins the most likely solution to Eq. (6.8) is that $M_X = M_Y = 0$, or that all the individual spins precess at random.

6.3. Bloch Equations

The net magnetization **M** produced by a group of spins is what is measured in magnetic resonance spectroscopy. One applies a high-frequency magnetic field, near radio or microwave frequencies, to a group of spins in a strong magnetic field H_0. If energy is exchanged between the spins and the high-frequency field, a magnetic resonance spectrum is observed. It is easy to see that energy will be exchanged only if the high-frequency field induces a net magnetization in the sample. This is an example of the forced precession of spins, and the necessary modifications of Eq. (6.8) were first developed by F. Bloch (1946).

The necessary high-frequency magnetic field must be applied in the XY plane. Since it is usually applied as an ac magnetic field produced by a coil of wire, if $2H_1$ is its magnitude along X and ω is its angular frequency,

$$H_X(\text{ac}) = 2H_1 \cos \omega t = H_1[\cos \omega t + \cos (-\omega t)]$$

$$H_Y(\text{ac}) = 0 = H_1[\sin \omega t + \sin (-\omega t)].$$

They are written this way to show that an ac field polarized along X can be represented by two rotating fields—one rotating in the positive ϕ direction and one in the negative. Since all the spins precess in the negative ϕ direction the only effective component of the ac field is that which is also rotating in the negative ϕ direction of Fig. 6.2, so that

$$H_X(\text{effective}) = H_1 \cos \omega t \tag{6.9a}$$

$$H_Y(\text{effective}) = -H_1 \sin \omega t. \tag{6.9b}$$

The spins will see a total magnetic field **H** which is the vector sum of Eq. (6.9) and the fixed Z-direction field which we now call H_0. To designate the free precession Larmor frequency we will now call it ω_0, but to avoid the negative sign it will be defined as

$$\omega_0 = 2\pi\nu_0 = \gamma H_0. \tag{6.10}$$

RELAXATION TIMES

Before the application of H_0 we shall assume that the spins were all at random. After the application of H_0, Boltzmann equilibrium requires that M_Z be no longer zero. A random orientation of spins in a field is one of maximum entropy but not of minimum free energy. Random spins that are suddenly placed in a magnetic field must slowly lose energy until the Boltzmann equilibrium is reached. The simplest possible form for this energy loss is that of a first-order decay. The time constant for that decay is called T_1, so, for M_Z, Bloch assumed that

$$\frac{dM_Z}{dt} = \frac{M_0 - M_Z}{T_1} \tag{6.11}$$

where M_0 is the equilibrium value for M_Z. The magnetic susceptibility of the spins is related to M_0 and

$$M_0 = H_0\chi(\text{spins}) \tag{6.12}$$

where χ(spins) can be calculated from Eq. (3.13).

There is a second kind of relaxation which is possible in a set of spins. Assume that a group of spins are precessing together to yield finite values for M_X and M_Y. If one waits long enough, these values must reduce to zero. This is because M_X and M_Y are zero for a truly random set of spins. The time constant for this second kind of relaxation is called T_2 and it is defined by

$$\frac{dM_X}{dt} = -\frac{M_X}{T_2} \tag{6.13a}$$

$$\frac{dM_Y}{dt} = -\frac{M_Y}{T_2}. \tag{6.13b}$$

Before we combine all these equations we must comment further on these relaxation times, for they are the heart of the Bloch equations. The question is: How does a group of free spins change its values for M_Z, M_X, and M_Y as Eqs. (6.11) and (6.13) require? The answer is that under most circumstances these changes can only take place by rather special kinds of processes. A change in M_X and M_Y will take place if all the spins do not precess at the same frequency. But according to Eq. (6.7) this is only possible if the magnetic field in the Z direction is not the same for all the spins. Since our spins are located in molecules that tumble or vibrate because of thermal excitation, one would expect a small fluctuating magnetic field to be present. The average value of these fluctuations is zero, but H will oscillate slightly above and below H_0. The effect of these fluctuations is to give a spin a distribution of Larmor frequencies and, as a result, an eventual randomization of M_X and M_Y toward the equilibrium zero value. The whole process is similar to runners on a circular track who speed up and slow down in a random manner. After several turns around the track they would be spread out all over the circle even if they had started together in one bunch.

The fluctuations of the magnetic field in the X and Y directions are responsible for the relaxation time T_1. To change M_Z energy must be exchanged between the spins and the molecules in which they are present. In solids these molecules form a lattice and T_1 is most commonly called the spin–lattice relaxation time. We shall see later that an oscillating magnetic field in the XY plane is just what is needed to change the value of M_Z. Both the amplitude and frequency of this oscillating magnetic field are important. In most cases, the thermal fluctuations are very fast and the magnitude of the fluctuating field at the correct frequency can be treated by statistical methods.

If H_X^* and H_Y^* are the magnitudes of these fluctuating fields, a simple formula for T_1 can be written. The result for the simplest spin case is

$$\frac{1}{T_1} = \gamma^2[(H_X^*)^2 + (H_Y^*)^2]\frac{\tau_c}{1 + \omega_0^2\tau_c^2}. \tag{6.14a}$$

The last term in this equation relates to the frequency of the fluctuations and introduces the correlation time τ_c. This is the characteristic time of the fluctuations and it will be discussed later.

The corresponding equation for T_2 has to include the fact that H_X^* and H_Y^* also affect T_2. When the spins change their energy, they also change the phase of their precession. As a result, the equation for the effect of fluctuations on T_2 is

$$\frac{1}{T_2} = \gamma^2\tau_c(H_Z^*)^2 + \frac{1}{2T_1}. \tag{6.14b}$$

Although these equations appear to be quite simple to use, it is often a very difficult task to calculate values for H_X^*, H_Y^*, and H_Z^* in many situations. This is particularly true of electron relaxation involving orbital angular momentum.

A careful analysis of these equations shows that $T_2 \leq T_1$. The conditions for the inequality obviously depend upon the value for τ_c. One important case comes about when τ_c is very small and $\omega_0\tau_c \ll 1$. Under these circumstances Eqs. (6.14*a*) and (6.14*b*) become very similar. For an isotropic fluctuating field we would expect that $(H_X^*)^2 = (H_Y^*)^2 = (H_Z^*)^2$. With this assumption and a short τ_c one has the limiting case where $T_1 = T_2$. We shall discuss this situation further for the proton resonance in liquid water.

The Bloch equations do not try to include an understanding of the source of T_1 and T_2, for they only treat them as adjustable parameters. In these equations T_1 and T_2 are called the longitudinal and transverse relaxation times, respectively. It is also common to call T_2 the spin–spin relaxation time and T_1 the spin–lattice relaxation time.

When Eqs. (6.8), (6.9), (6.11), and (6.13) are combined, the resultant Bloch differential equations are

$$\frac{dM_X}{dt} = \gamma(M_Y H_0 + M_Z H_1 \sin \omega t) - \frac{M_X}{T_2} \tag{6.15a}$$

$$\frac{dM_Y}{dt} = -\gamma(M_X H_0 - M_Z H_1 \cos \omega t) - \frac{M_Y}{T_2} \tag{6.15b}$$

$$\frac{dM_Z}{dt} = -\gamma(M_X H_1 \sin \omega t + M_Y H_1 \cos \omega t) + \frac{M_0 - M_Z}{T_1}. \tag{6.15c}$$

There are many possible sets of solutions to these equations. Each set of solutions depends upon the boundary conditions placed upon these differential equations. The most common type of magnetic resonance spectroscopy applies a uniform field H_1 and then slowly varies either its frequency ω or the Z-axis-field H_0. The solutions appropriate to these conditions are discussed in Section 6.4.

6.4. *Steady State Susceptibilities*

The Bloch equations are most easily solved if we express them in terms of magnetizations that rotate along with the effective transverse field defined in Eq. (6.9). The magnetizations u and v are illustrated in Fig. 6.3, so that

$$M_X = u \cos \omega t - v \sin \omega t \tag{6.16a}$$

$$M_Y = -(u \sin \omega t + v \cos \omega t). \tag{6.16b}$$

When Eq. (6.15) is transformed to the u and v magnetizations,

$$\frac{du}{dt} = -\Delta\omega\, v - \frac{u}{T_2} \tag{6.17a}$$

$$\frac{dv}{dt} = \Delta\omega\, u - \frac{v}{T_2} - \gamma H_1 M_Z \tag{6.17b}$$

$$\frac{dM_Z}{dt} = \gamma H_1 v + \frac{M_0 - M_Z}{T_1} \tag{6.17c}$$

where $\Delta\omega = \gamma H_0 - \omega = \omega_0 - \omega$.

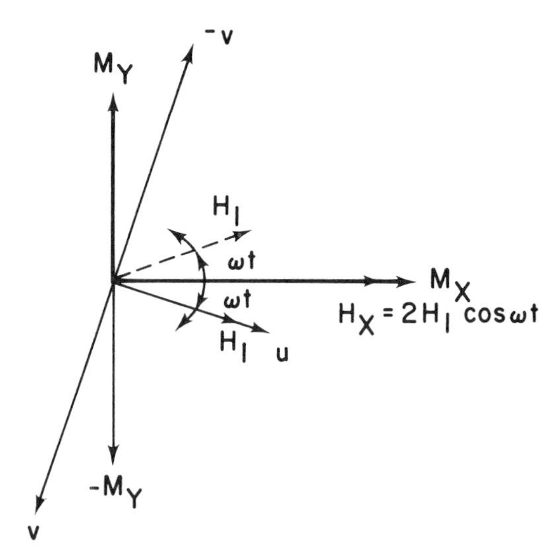

Figure 6.3. *Relation between stationary and rotating magnetizations. The Z axis points out of the plane of the paper. The magnetization rotates along with u and v and with the component of H_X which rotates clockwise on this figure. The dotted component of H_X is not effective, for it does not follow the magnetization.*

The steady-state solutions to these equations would be appropriate when H_1, H_0, or ω only change very slowly with time. If we assume a steady state,

$$\frac{du}{dt} = \frac{dv}{dt} = \frac{dM_Z}{dt} = 0,$$

and we find for Eq. (6.17) that

$$u = \frac{\gamma H_1 \, \Delta\omega \, T_2^2 M_0}{1 + (\Delta\omega)^2 T_2^2 + \gamma^2 H_1^2 T_1 T_2} \tag{6.18a}$$

$$v = \frac{-\gamma H_1 T_2 M_0}{1 + (\Delta\omega)^2 T_2^2 + \gamma^2 H_1^2 T_1 T_2} \tag{6.18b}$$

$$M_Z = \frac{[1 + (\Delta\omega)^2 T_2^2] M_0}{1 + (\Delta\omega)^2 T_2^2 + \gamma^2 H_1^2 T_1 T_2} \, . \tag{6.18c}$$

These three magnetizations can be used to define three susceptibilities. The ordinary, or longitudinal, susceptibility is that established along the

Z axis by the field H_0 and

$$\chi(\text{long.}) = \frac{M_Z}{H_0}\,. \tag{6.19}$$

There are two components to the transverse susceptibility One component, called χ', is in phase with H_X and rotates as u does along with the effective component H_1, so that†

$$\chi' = \frac{u}{2H_1}\,. \tag{6.20a}$$

The second component, called χ'', is 90° out of phase with H_X and rotates along with $-v$, so that†

$$\chi'' = -\frac{v}{2H_1}\,. \tag{6.20b}$$

These two components, χ' and χ'', can be called the real and imaginary parts, respectively, of a complex transverse susceptibility.

An examination of Eq. (6.18) shows that there are two limits to the solution of the Bloch equations even in the steady state. If we have

$$\gamma^2 H_1^2 T_1 T_2 \ll 1, \tag{6.21}$$

our equations are all simplified. This is quite possible in a given experiment since H_1 can be adjusted by the experimenter. It may be difficult to observe signals with such small values of H_1, particularly if $T_1 T_2$ is large, but like the steady state it is a limiting approximation for any experiment.

LOW POWER RESULTS

At low applied ac power Eq. (6.21) is accurate. In this case the susceptibilities have the form

$$\chi(\text{long.}) = \frac{M_0}{H_0} = \chi_0 \tag{6.22a}$$

$$\chi' = \frac{1}{2}\chi_0\omega_0\left[\frac{(\Delta\omega)T_2^2}{1 + (\Delta\omega)^2 T_2^2}\right] \tag{6.22b}$$

$$\chi'' = \frac{1}{2}\chi_0\omega_0\left[\frac{T_2}{1 + (\Delta\omega)^2 T_2^2}\right]. \tag{6.22c}$$

† We divide by $2H_1$ because we defined $H_X = 2H_1 \cos \omega t$. Some authors define H_X differently.

A plot of χ'' in Fig. 6.4 shows that when $\Delta\omega = 1/T_2$ it has dropped to one-half its maximum value at $\Delta\omega = 0$. It also shows that χ' has its maximum and minimum values when $\Delta\omega = \pm(1/T_2)$.

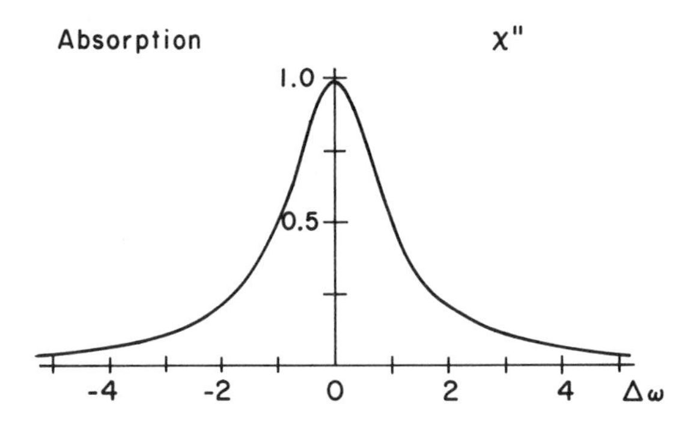

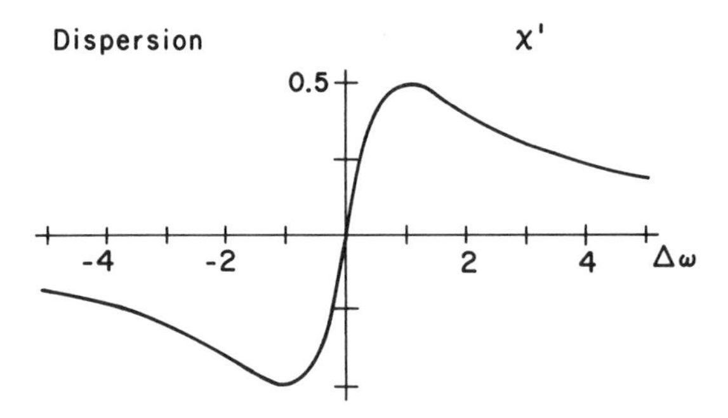

Figure 6.4. *Low power absorption and dispersion:* χ'' *and* χ' *are plotted in units of* $\frac{1}{2}\chi_0\omega_0T_2$ *vs.* $\Delta\omega$ *in units of* $1/T_2$.

At low power, χ(long.) is undisturbed from its equilibrium value, called χ_0. The transverse component χ'', which is out of phase with the ac magnetic field, is maximum when ω equals the Larmor frequency. This component is related to the absorption of ac power by the system of spins and it forms the line-shape function for the absorption signals in many magnetic resonance spectrometers. The specific function in the square brackets in Eq. (6.22*c*) is called the Lorentz line shape, in honor of H. A. Lorentz, who quite early in the history of spectroscopy showed that classical damped oscillators should yield this line-shape function.

The in-phase susceptibility χ' is called the dispersion. It has the characteristics of slowing down the applied electromagnetic radiation on one side of resonance and speeding it up on the other. A simple NMR spectrometer is illustrated in Fig. 1.9. If the coil is always driven at its resonant frequency and the magnetic field is slowly swept through the spin's Larmor resonance, only χ'' will be detected. However, if the coil is operated off its resonance frequency, the detector will receive signals that are due to changes both in the absorption of the sample and in it dispersion. A marginal oscillator using the sample coil in the frequency-determining circuit is a simple and sensitive spectrometer for the determination of χ''. Most commercial NMR spectrometers of course, have more sophisticated methods to give signals that measure the absorption of the spins, or to be more exact, χ''.

POWER SATURATION

If H_1 is increased so that Eq. (6.21) is no longer satisfied, χ(long.) becomes less than χ_0. This is called a condition of power saturation. Inspection of Eq. (6.18*c*) shows that M_Z is reduced in value at sufficiently high values of H_1.

The primary effect of H_1 is to transfer energy into the spin system when ω is close to resonance. In Eq. (6.11) we assumed that the spin system can only lose energy at a finite rate. If H_1 is too large, the spin system becomes "saturated" with energy that it cannot lose fast enough. This saturation can be clearly seen from Fig. 6.1, where absorption of a quantum of energy excites one spin in the system to its higher energy state. If it cannot rapidly lose this energy, by a spontaneous process, the Boltzmann equilibrium will be disturbed and χ(long.) will be decreased below its equilibrium value. Some workers like to refer to a "spin temperature," where they use Boltzmann's relation to define their "temperature." In this picture, complete power saturation has equal populations in the upper and lower spin levels and is one of infinite spin temperature. The relaxation time T_1 is, of course, the time constant for spontaneous changes in the spin temperature, and most of the methods that are used to measure T_1 are essentially based upon its measurement.

6.5. Non-Steady-State Conditions

If rapid changes are made in H_1 or ω, then the steady-state assumptions for the rotating magnetization are no longer valid. A result more general even than the Bloch equations can be obtained by starting with Eq.

(6.8), so that

$$\frac{d(\mathbf{M}\cdot\mathbf{M})}{dt} = \gamma\mathbf{M}\times\mathbf{H}\cdot\mathbf{M}$$
$$= \gamma\mathbf{H}\cdot\mathbf{M}\times\mathbf{M} = 0. \tag{6.23}$$

One can now see that if H_1 is rapidly changed the magnitude of the magnetization is conserved. One finds that the magnetization is a vector that can be oriented in various directions by properly selected values of the applied magnetic fields.

One experiment is to apply an H_1 large enough to produce saturation and then to sweep through resonance in a time less than T_2. This has been called "fast passage" by Bloch. This can result in a rotation of the magnetization vector, so that the transverse magnetization becomes equal to M_0 during passage and the longitudinal magnetization becomes equal to $-M_0$ after passage. A second fast passage can then show how much M_Z has decayed back to its equilibrium value of $+M_0$ and thereby measure T_1.

One can observe a series of "wiggles" immediately following the absorption signal in high-resolution NMR spectrometers when H_1 is kept small and the absorptions are swept through in a time less than T_2. These "wiggles" are illustrated in Fig. 6.5. They result from the difference

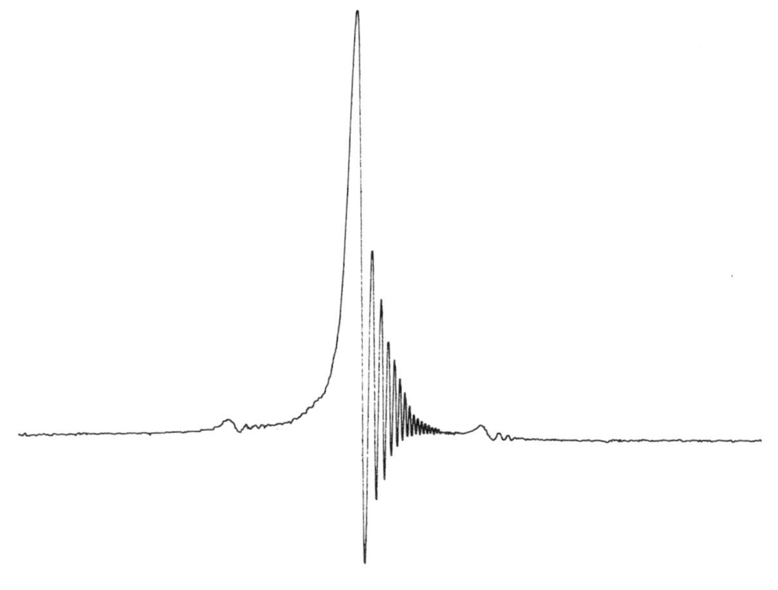

Figure 6.5. *Rapidly swept proton NMR spectrum of tetramethylsilane. The ringing is only found following the passage through resonance if times shorter than T_2 are utilized. The weak doublet comes from an isotope of Si.*

in frequency between ω and the magnetization, which tries to continue to precess at γH_0. The magnetization will continue to precess until it decays according to T_2, or until the spins are dephased by magnet inhomogeneities. The "wiggles" are used as a practical test of magnet homogeneity.

A very important technique for nuclear resonance utilizes a series of pulsed H_1 signals. By choosing the amplitude and spacings of these pulsed H_1 values "spin echoes" can be observed. These echoes can be used to measure T_1, T_2, and all the other parameters that characterize a system of spins. The spin-echo technique is a particularly powerful technique in those systems where H_1 values can be applied which lead to saturation and rotation of the magnetization vector into the XY plane.

6.6. *Effect of Fluctuations*

Let us suppose that a set of spins can, even at constant H_0, have two possible Larmor frequencies, which we will call $\omega_0 + \delta\omega$ and $\omega_0 - \delta\omega$. If there is a factor in the system that makes the spins rapidly fluctuate between these two states, one will observe a single resonance which is centered about ω_0. The fluctuations will contribute to the width of the observed resonance or, in other words, to $1/T_2$. Simple statistics can calculate this width.

If the time between fluctuations is called τ_c and one waits for a total time such that n fluctuations have taken place, the root-mean-square (rms) phase difference between the precession of all the spins is

$$(\delta\phi^2)_{\text{av}}^{1/2} = \sqrt{n}\,\tau_c\,\delta\omega \tag{6.24a}$$

where $\tau_c\,\delta\omega$ is the phase difference that builds up before a new fluctuation takes place. If the fluctuations are much shorter than T_2, we can wait for the time T_2 for our n fluctuations, or then $n = T_2/\tau_c$ and

$$(\delta\phi^2)_{\text{av}}^{1/2} = \sqrt{\frac{T_2}{\tau_c}}\,\tau_c\,\delta\omega = \delta\omega\sqrt{T_2\tau_c}\,. \tag{6.24b}$$

In the Bloch equations the time T_2 is defined as the time constant for the decay of the transverse magnetization. This is the same as the time for the rms value of the phase angle of the precession of all the spins to reach about 1 radian. So if we wait for the time T_2 in Eq. (6.24*b*),

$$(\delta\phi^2)_{\text{av}}^{1/2} \approx 1$$

or

$$\frac{1}{T_2} \approx \tau_c(\delta\omega)^2. \tag{6.25}$$

This result is identical to the first term in Eq. (6.14*b*).

One example of the use of Eq. (6.25) in NMR would be if a nucleus rapidly jumped between two molecular positions, each with a different chemical shift. For some systems T_2 values close to 1 second are possible, so that many chemical exchange reactions can be much faster than this. In liquids the tumbling of molecules tends to average out all anisotropic contributions to the Larmor frequency. In this case Eq. (6.25) shows how fast molecular tumbling can reduce the resonance line widths in liquids over that observed in solids. There are many examples in magnetic resonance where Eq. (6.25) can show, in a qualitative way, how much fast fluctuations can be expected to narrow the resonance signals.

CORRELATION TIMES

The time τ_c is called the correlation time. It is a direct measure of the speed of the fluctuations. In systems involving chemical exchange it is the lifetime of a single species. In the removal of anisotropic interactions in liquids it is the time for about 1 radian of molecular rotation. In their pioneering work Bloembergen, Purcell, and Pound (1948) developed a reasonable mathematical technique for introducing correlation times into magnetic resonance. Their immediate problem was the averaging in liquid water of the magnetic field at one of the protons due to the magnetic moment of the other proton. As Fig. 5.5 shows, the field in a molecule produced by an aligned magnetic moment is averaged to zero by molecular tumbling. BPP, as their work is commonly called, shows how changes in a correlation time affect both T_1 and T_2. A graph of these results is shown in Fig. 6.6, and it follows the pattern shown in Eqs. (6.14*a*) and (6.14*b*).

The observed T_1 and T_2 values in magnetic resonance vary a great deal depending upon the correlation times, upon the magnitude of the broadening interaction, and upon ω_0. A list of representative values for T_1 and T_2 for various systems is given in Table 6.1. One can see from this table that the T_1 values and predicted line widths can vary over an enormous range. All the systems in this table give magnetic resonance signals that can be directly observed, except for Ni^{+2} in water. For this, the predicted line width is essentially greater than any of the commonly applied fields.

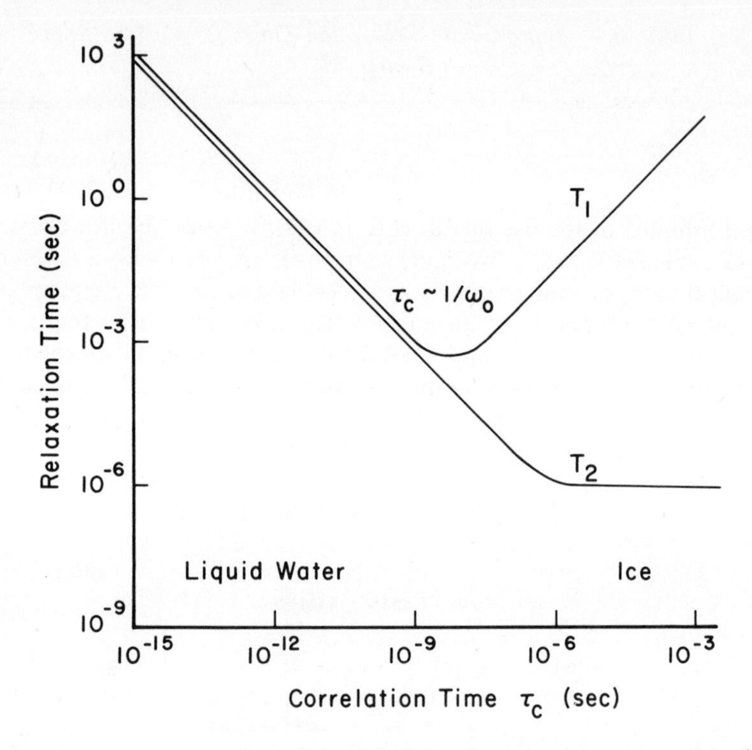

Figure 6.6 *Predicted behavior of T_1 and T_2 for the protons in water: the calculation by BPP for $\omega_0 \approx 2 \times 10^8$ rad/sec. At room temperature, the observed $T_1 \approx T_2 \approx 1$ sec corresponds to $\tau_c \approx 10^{-12}$ sec.*

The value for T_1 in LiF is probably determined by very small quantities of paramagnetic impurities, for, as Fig. 6.6 shows, solids with correlation times longer than 1 second can have very long values for T_1. For Li^+ in water we find $T_1 = T_2$ and both are large, as the left side of Fig. 6.6 predicts for liquids.

For Cl^- the values of T_1 and T_2 are much shorter than for Li^+. This is due to nuclear quadrupole relaxation. Those nuclei with $I \geq \frac{1}{2}$ can possess electric nuclear quadrupole moments. In situations where the molecular electric field at the nucleus has a symmetry lower than tetrahedral, molecular rotation can lead to nuclear rotation. This is illustrated diagrammatically in Fig. 6.7. Although ^7Li has a quadrupole moment, there are only small electric field gradients at this nucleus in most of its compounds. This is not true for ^{35}Cl, and liquid $SiCl_4$ strongly shows the effects of nuclear quadrupole relaxation.

Most of the relaxation times for electrons are shorter than they are for nuclei. This is a result of the quenched orbital magnetic moment and

Table 6.1. *Approximate Relaxation Times for Various Spin Systems*[a]

System	T_1 (sec)	T_2 (sec)	Predicted line width[b] (gauss)
^{7}Li in LiF(s)	10^3	$\sim 10^{-5}$	~ 10
^{7}Li as Li^+ in H_2O(l)	3	3	0.00007
^{35}Cl as Cl^- in H_2O(l)	2×10^{-2}	2×10^{-2}	0.04
^{1}H in H_2O(s)	10	10^{-5}	10
^{1}H in H_2O(l)	2	2	0.00004
^{35}Cl in $SiCl_4$(l)	5×10^{-5}	5×10^{-5}	15
e^- in NH_3(l)	3×10^{-6}	3×10^{-6}	0.04
e^- as $(SO_3)_2NO^{-2}$ in H_2O(l)	$\sim 10^{-6}$	3×10^{-7}	0.4
e^- as Mn^{+2} in H_2O(l)	$\sim 10^{-8}$	3×10^{-9}	40
e^- as Ni^{+2} in H_2O(l)	4×10^{-12}	4×10^{-12}	30,000

[a] Near or just below room temperature. For nuclei at fields close to 10,000 gauss and for electrons close to 3000 gauss.
[b] Line width defined as $2/\gamma T_2$.

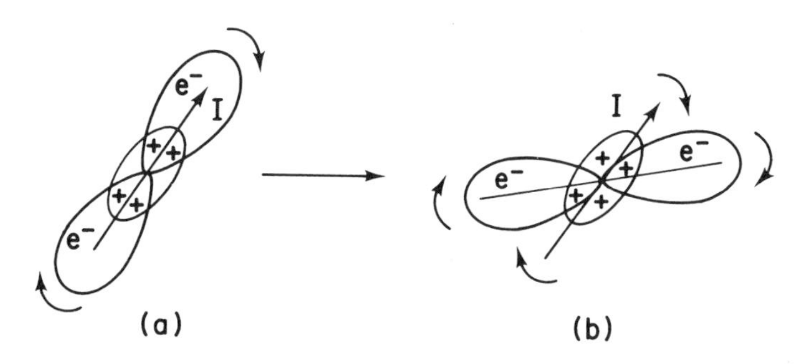

Figure 6.7. *How a nuclear quadrupole moment would be oriented by a rotating electron distribution. The nucleus is prolate ($Q > 0$) and the electron is in a p orbital. The lower electrostatic energy in orientation (a) tends to reorient the nucleus in (b).*

spin–orbit interaction. In paramagnetic molecules the small residual orbital moment can generate relatively large fluctuating magnetic fields and these fields produce short values for the relaxation times of electrons. The two common circumstances that yield fairly long relaxation times for electrons are very well quenched orbital moments or complexes derived from *S*-state ions with no orbital moments. The paramagnetic ion $(SO_3)_2NO^{-2}$ appears to be an example of highly quenched orbital angular momenta and Mn^{+2} in water is derived from a ^6S-state ion. It also seems to be true that the solvated electron dissolved in liquid ammonia is related to an *S* state.

In Chapters 7 and 8 we will discuss some spin systems in greater detail. In particular, we will find that for most systems there are energies in addition to the Zeeman energy that must be considered. Most work in magnetic resonance spectroscopy is concerned with the fine structure associated with these additional sources of energy. Thus once a spectrum can be observed, much attention must be paid to the complexities that are often readily observed.

References

A. Abragram, *The Principles of Nuclear Magnetism*, Oxford University Press, Inc., New York (1961). Rather hard to read, but nearly the last word on spin relaxation in solids, liquids, and gases.

E. R. Andrew, *Nuclear Magnetic Resonance*, Cambridge University Press, New York (1956). An old favorite for fundamentals.

N. Bloembergen, *Nuclear Magnetic Relaxation*, W. A. Benjamin, Inc., Menlo Park, Calif. (1961); N. Bloembergen, E. M. Purcell, and R. V. Pound, *Phys. Rev.*, **73,** 679 (1948). The classic thesis and paper on nuclear relaxation in solution.

A. Carrington and A. D. McLachlan, *Introduction to Magnetic Resonance*, Harper & Row, Publishers, New York (1967). The best general introduction to NMR and EPR.

E. L. Hahn, "Nuclear Induction," *Phys. Today*, **6,** No. 11 (1953), p. 4. A short elementary introduction to "spin echoes."

C. P. Slichter, *Principles of Magnetic Resonance*, Harper & Row, Publishers, New York (1963). All theory on a fairly advanced level.

Problems

6.1. Calculate ω_0 for protons at a field of 20,000 gauss. Compare this with the vibrational frequency in H_2 where $\tilde{\nu}(\text{vib}) = 3900\ \text{cm}^{-1}$ and the rotational frequency for $J = 1$ where $\bar{B} = 60\ \text{cm}^{-1}$.

6.2. Which nucleus in Section A.3 has the largest value for γ? Is it necessarily the one with the largest value for μ_I? What is the ratio of γ(electron) for $g = 2$ and γ(proton)? Ignore the fact that γ can be either positive or negative.

6.3. The equation for χ_0 and M_0 is given in explicit form in Chapter 3 for a mole of electron spins. Derive the same equation for 1 mole of nuclear spins.

6.4.* Calculate H_1 in gauss for the condition that $\gamma^2 H_1^2 T_1 T_2 = 1$ for the protons in liquid water and for ^{35}Cl in liquid $SiCl_4$.

6.5. A common way to make the proton signal in water less subject to power saturation is to dissolve paramagnetic ions in it. Most NMR magnetometers use water "doped" with Cu^{+2} or Mn^{+2}. Calculate the approximate line width in gauss for 0.1 M Cu^{+2} where $T_1 \approx T_2 = 10^{-2}$ sec. How much larger a value of H_1 can be used in this "doped" water compared with pure water before saturation becomes important?

6.6. Solve for M_X and M_Y using Eqs. (6.16) and (6.18). Show that both these components of the magnetization oscillate in time as predicted by Fig. 6.3.

6.7.* The power absorbed during resonance is most easily calculated using

$$\frac{dE}{dt} = P = -\mathbf{M} \cdot \frac{d\mathbf{H}}{dt}$$

in the rotating coordinate system. Show that since

$$\mathbf{H} = H_0\mathbf{k} + H_1 \cos \omega t\mathbf{i} - H_1 \sin \omega t\mathbf{j}$$

that $d\mathbf{H}/dt$ rotates along with v and has the value ωH_1. Solve for P and show that it equals $2H_1^2\omega\chi''$.

6.8. Since the absorption signal is proportional to χ'', show that at high power the absorption signal is broadened. That is, when Eq. (6.21) is not satisfied, the half-maximum point along χ'' comes when $\Delta\omega > 1/T_2$.

6.9. Express χ'' in a form suitable for spectroscopy where the frequency is fixed and the Z-axis field is varied. Take the fixed frequency to be ω_0 and the field at exact resonance to be H_0. Then the field away from exact resonance is ΔH where $\Delta H = \Delta\omega/\gamma$. In other words, transform $\chi''(\Delta\omega)$ into $\chi''(\Delta H)$. If Eq. (6.21) is satisfied, what is ΔH when χ'' has dropped to one-half its maximum value?

6.10. One of the unsatisfactory features of the Bloch equations for systems with large values for $1/T_2$ is that they do not predict zero power absorption when $\omega = 0$. What value does Eq. (6.22*c*) give for χ'' when $\omega = 0$? What assumption in the Bloch equations do you think would have to be modified to get rid of this unsatisfactory feature? An answer to this question is found in Abragam on page 53.

6.11. Assume that the protons in H_2O are 1.5×10^{-8} cm apart. Estimate a value for the H^* produced by one proton at the site of the other when H_2O tumbles in solution using the formulas given in Fig. 5.5. Relate T_1 to H^* and τ_c in Eq. (6.14*a*) under the conditions that $\omega_0\tau_c \ll 1$ and $T_1 = T_2$. Use your results to check Fig. 6.6 for liquid water when $\tau_c = 10^{-12}$ sec.

Solutions:

6.4. $H_1 = \dfrac{1}{\gamma(T_1T_2)^{1/2}}$.

For protons,

$$\gamma = \frac{4\pi\mu_I\mu_N}{h} = 2.67 \times 10^4 \text{ sec}^{-1} \text{ G}^{-1};$$

and with $T_1 = T_2 = 2$ sec,

$$H_1 = (2 \times 2.67 \times 10^4)^{-1} = 2.0 \times 10^{-5} \text{ gauss.}$$

For ^{35}Cl,

$$\gamma = \frac{4\pi\mu_I\mu_N}{3h} = 2.60 \times 10^3 \text{ sec}^{-1} \text{ G}^{-1};$$

and with $T_1 = T_2 = 5 \times 10^{-5}$ sec,

$$H_1 = (5 \times 2.6 \times 10^{-2})^{-1} = 8 \text{ gauss.}$$

6.7. $\dfrac{d\mathbf{H}}{dt} = -\omega H_1(\sin \omega t\mathbf{i} + \cos \omega t\mathbf{j})$.

From Fig. 6.3 one can see that $-v$ also rotates according to $\sin \omega t\mathbf{i} + \cos \omega t\mathbf{j}$. Under these circumstances

$$\mathbf{M} \cdot \frac{d\mathbf{H}}{dt} = v\omega H_1.$$

With Eq. (6.20*b*) one obtains

$$P = \text{power absorbed} = 2H_1^2\omega\chi''.$$

This shows that the power absorbed follows χ''. Since the power applied is proportional to H_1^2, it gives a fixed absorption coefficient as long as χ'' is independent of H_1.

7

Nuclear Magnetic Resonance

IN CHAPTER 6 WE SAW that NMR can have very narrow absorption lines when certain nuclei are examined in liquid systems. Work on these systems is usually termed "high resolution." In solids or for some nuclei in solution much broader lines are observed. Work of this type is usually called "broad line." Although there is a complete continuity of possible line widths in NMR, this is a convenient separation, because two different kinds of instruments are used for these two cases. High-resolution spectrometers must have a very uniform and stable magnetic field. Sample sizes are kept small, and the sample tubes are always spun in these instruments. Although all these efforts are devoted to obtaining the narrowest possible lines in a high-resolution spectrometer, the observed line widths are usually still determined by instrumental factors. Most spectrometers are also designed to operate at as high an H_0 value as is reasonable, for the sensitivity of NMR instruments increases with H_0.

Narrow lines are obtained in diamagnetic liquid systems when the absorbing nucleus has no nuclear quadrupole interaction. This is always true when $I = \frac{1}{2}$ and these nuclei form the basis for high-resolution NMR. The most popular of these nuclei are ^{1}H, ^{13}C, ^{15}N, ^{19}F, ^{29}Si, and ^{31}P. At times nuclei with $I > \frac{1}{2}$ give fairly narrow NMR lines and ^{2}H, ^{7}Li, ^{14}N, ^{17}O, and several other nuclei can also be examined under conditions of fairly high resolution.

In broad-line NMR, sample sizes can be several cubic centimeters, and line widths have fundamental importance since they are not determined by instrumental factors. Much of the work in broad-line NMR is concerned with the analysis of line broadening. In many systems one finds that this broadening is related to the chemistry of the system under study. Several books have been written about the application of proton NMR to the study of organic molecules. Such work is an important use of high-resolution NMR but it does not illustrate the entire range of studies that are possible with NMR. In this chapter we shall primarily pick examples from the physical and inorganic literature, for fewer summaries of this area of NMR are to be found at the elementary level.

7.1. High–Resolution Hamiltonian

In liquids the rapid molecular tumbling can remove all the terms in the Hamiltonian that depend upon the relative orientation of a molecule and the applied magnetic field. Although these anisotropic terms may still contribute to the line widths, as explained in Chapter 6, the basic Hamiltonian in liquids is isotropic. For a single nucleus the Zeeman part of the Hamiltonian is

$$\mathsf{H}(\text{Zeeman}) = -\boldsymbol{\mu} \cdot \mathbf{H} = -\frac{\gamma h}{2\pi}(1 - \sigma)H_0 \mathsf{I}_Z, \tag{7.1a}$$

where the shielding σ has been introduced in accordance with Eq. (5.15). The Larmor frequency ν_0 is often used to simplify Eq. (7.1*a*), and if h is factored out, then

$$h^{-1}\mathsf{H}(\text{Zeeman}) = -\nu_0(1 - \sigma)\mathsf{I}_Z \tag{7.1b}$$

where this equation is now in units of frequency.

If there were no interactions at all between nuclei, Eq. (7.1*b*) could be summed over all the nuclei in a molecule and separate Zeeman energies obtained for each nucleus. However, even in a tumbling molecule there is an interaction between nuclei that does not average to zero. We previously saw that direct dipole–dipole interaction should average to zero, but the electrons in the molecule yield an indirect interaction whose isotropic value is not zero. This interaction is a spin–spin coupling and can be written as a dot product. For two nuclei i and j

$$h^{-1}\mathsf{H}(\text{spin–spin}) = J_{ij}\mathbf{I}_i \cdot \mathbf{I}_j \tag{7.2}$$

where J_{ij} is a spin–spin coupling constant in frequency units. For several nuclei, Eqs. (7.1b) and (7.2) can be combined, so that the total nuclear spin Hamiltonian is

$$h^{-1}\mathsf{H}(\text{spin}) = -\sum_i \nu_{0i}(1 - \sigma_i)\mathsf{I}_{Zi} + \sum_{i<j} J_{ij}\mathbf{I}_i \cdot \mathbf{I}_j. \tag{7.3}$$

SPIN–SPIN INTERACTION

Before we attempt to use Eq. (7.3) we should know more about the J_{ij} values, which only have magnitudes between 1 and 1000 cycles per second (hertz, Hz). The Fermi contact interaction between electrons and nuclei introduced in Eq. (2.42) is a form of spin–spin interaction but one of different origin. The Fermi contact interaction arises from the fact that electrons and nuclei can occupy the same space, but this is not true of two nuclei. The nuclear spin–spin interaction is produced by way of the electrons and it is properly called electron-coupled nuclear spin–spin interaction. Figure 7.1 gives a picture of the method of electron coupling for the two protons in H_2.

Each of the two nuclei in Fig. 7.1 interacts with the electrons that surround it. The most important intraction is most often via the Fermi contact term. Since A is positive in Eq. (2.42) when the left nucleus points "up," a very slightly lower energy will be obtained if it is surrounded by an electron that points "down," as indicated in Fig. 7.1.

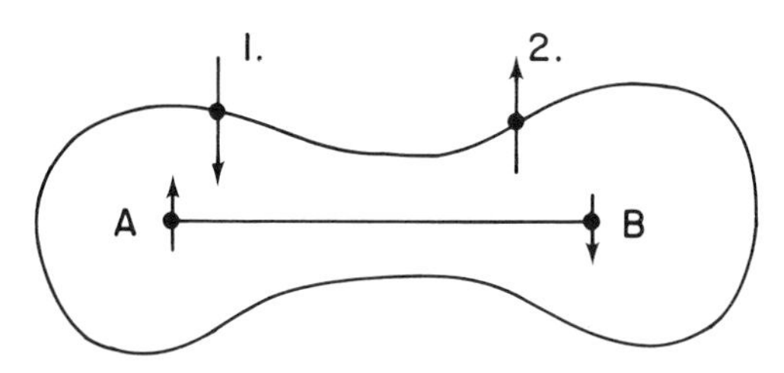

Figure 7.1. *Electron-coupled nuclear spin–spin interaction. Fermi contact interaction between nuclei A and B and electrons* 1 *and* 2 *produces a lower energy with* $\mathbf{I}_a \cdot \mathbf{I}_b$ *negative in an electron-pair bond.*

Table 7.1. *Approximate Spin–Spin Coupling Values*

Bonding situation	J (Hz)
H—H	280
>C(H)H	10 (H, H) 200 (^{13}C, H)
>C(H)F	50 (H, F) 200 (^{13}C, F)
>Si(H)H	10 (H, H) 200 (^{29}Si, H)
>P—F	1000 (P, F)
H—C—C—H	5 (H, H)
H—C=C—H	10 (H, H)
H—C=C(—H)	20 (H, H)

In an electron-pair bond this leaves the electron on the right nucleus pointing "up." Again if we use Fermi contact interaction, the lower energy will be obtained if the right nucleus now points "down." The net result of these considerations is that the spin–spin coupling constant J for the protons in H_2 should be positive in value. By experiment and careful calculation one can show that $J = +280$ Hz in H_2. Table 7.1 lists some approximate J values observed in certain bonding situations.

AB PROBLEM

If we first consider a molecule containing only two nuclei, A and B, each with the same Larmor frequency but with different shielding parameters, there can be only a single spin–spin coupling constant, so that Eq. (7.3) becomes

$$h^{-1}\mathsf{H}(AB) = -\nu_0[(1 - \sigma_a)\mathsf{I}_{Za} + (1 - \sigma_b)\mathsf{I}_{Zb}] + J\mathbf{I}_a \cdot \mathbf{I}_b. \tag{7.4}$$

At fields of a few thousand gauss ν_0 is much larger than any J value, but we shall see that the important aspect of Eq. (7.4) is how J compares with the difference between $\nu_0\sigma_a$ and $\nu_0\sigma_b$.

In H_2 and in similar molecules σ_a and σ_b are equal. This forms a special case of the AB problem called the A_2 case. The solutions for the Hamiltonian in this case can be easily formulated. In the vector model, the A_2 case corresponds to the exactly equal precession of $\mathbf{I}_a$ and $\mathbf{I}_b$. A coupled representation will give exact solutions, and if we form $\mathbf{F}$ by

$$\mathbf{I}_a + \mathbf{I}_b = \mathbf{F}, \tag{7.5}$$

Eq. (7.4) can be simplified when $\sigma_a = \sigma_b = \sigma$ as

$$h^{-1}\mathsf{H}(A_2) = -\nu_0(1 - \sigma)\mathsf{F}_Z + J\mathbf{I}_a \cdot \mathbf{I}_b. \tag{7.6a}$$

Since $\mathbf{I}_a^2$, $\mathbf{I}_b^2$, $\mathbf{F}^2$, and F_Z commute, we can quantize them following the methods of Chapter 2 and directly write down the energy levels. The dot product follows from Eq. (2.19) and with $I_a = I_b = I$,

$$h^{-1}\epsilon = -\nu_0(1 - \sigma)M_F + \frac{J}{2}[F(F + 1) - 2I(I + 1)]. \tag{7.6b}$$

To simplify matters we can take $I = \frac{1}{2}$, and Eq. (7.6*b*) gives four levels: $F = 0$, $M_F = 0$, and $F = 1$, $M_F = 0, \pm 1$. The solutions to the A_2 case for $I = \frac{1}{2}$ are shown in Fig. 7.2.

The general solution to Eq. (7.4) can also be obtained from the results of Chapter 2. There is a complete correspondence between Eq. (2.46) for the Zeeman effect in the hydrogen atom and Eq. (7.4) for two nuclei. The solutions for Eq. (2.46) are given for the $F = 1$, $M_F = \pm 1$ by Eq. (2.49) and for $F = 0$, $M_F = 0$, and $F = 1$, $M_F = 0$ by Eq. (2.51). For

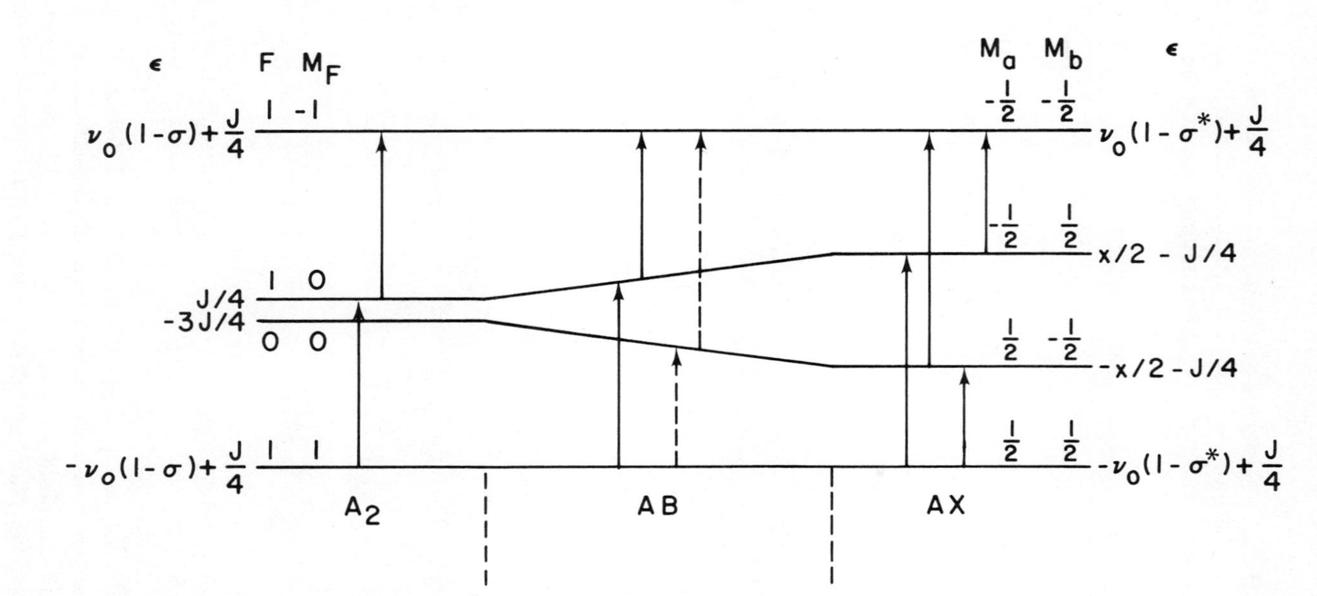

Figure 7.2. *Correlation diagram for the AB problem for spin* $= \frac{1}{2}$. *The energy levels are drawn so that* ν_0 *and J are constant. The AX case has* $\sigma^* = \frac{1}{2}(\sigma_a + \sigma_b) = \sigma(A_2)$ *and with* $\sigma_a < \sigma_b$. *Not drawn to scale since ordinarily* $\nu_0 \approx 10^6 J$.

$I = \frac{1}{2}$, if we use these results,

$$h^{-1}\epsilon(F = 1, M_F = \pm 1) = h^{-1}\epsilon(M_a = M_b = \pm \tfrac{1}{2})$$

$$= \frac{J}{4} \mp \nu_0\left(1 - \frac{\sigma_a + \sigma_b}{2}\right) \tag{7.7a}$$

$$h^{-1}\epsilon(M_F = 0) = -\frac{J}{4} \pm \frac{1}{2}(J^2 + x^2)^{1/2} \tag{7.7b}$$

where $x = \nu_0(\sigma_b - \sigma_a)$. These results are also shown in Fig. 7.2.

If x^2 is large compared to J^2 in Eq. (7.7*b*), then another limiting case is possible. This corresponds to the Paschen–Back effect where a large difference in the two precessional frequencies uncouples the angular momenta. This is represented by the vector-model diagram in Fig. 2.1*b*. In NMR this is called the AX case, where X designates a nucleus quite different from A. If we neglect J^2 in the second term in Eq. (7.7*b*), we obtain the AX limiting case,

$$h^{-1}\epsilon(M_F = 0, x^2 \gg J^2) = -\frac{J}{4} \pm \frac{x}{2}. \tag{7.7c}$$

These energies for the AX case are also shown in Fig. 7.2.

A detailed account of the allowed transitions is given in Section A.5, but it is easy to obtain a qualitative picture. The general dipole selection rules for angular momentum as applied to the A_2 problem are

$$\Delta F = 0 \qquad \Delta M_F = \pm 1 \tag{7.8a}$$

where we assume, as we stated in Chapter 6, that H_1 is perpendicular to H_0. In the AX case the quantum numbers M_a and M_b are appropriate and the selection rules are

$$\Delta M_a = \pm 1, \Delta M_b = 0 \qquad \text{or} \qquad \Delta M_b = \pm 1, \Delta M_a = 0. \tag{7.8b}$$

The absorption transitions allowed by these selection rules are also shown in Fig. 7.2. In the AB case two of the transitions are allowed both for A_2 and AX and two are allowed only for AX. These second transitions are shown as dotted in Fig. 7.2 since their relative intensity varies rapidly with the magnitudes of J and x.

In Fig. 7.3 we show the exact relative intensities and splittings expected for the AB problem with three x values and a fixed J. When $x = 0$ we

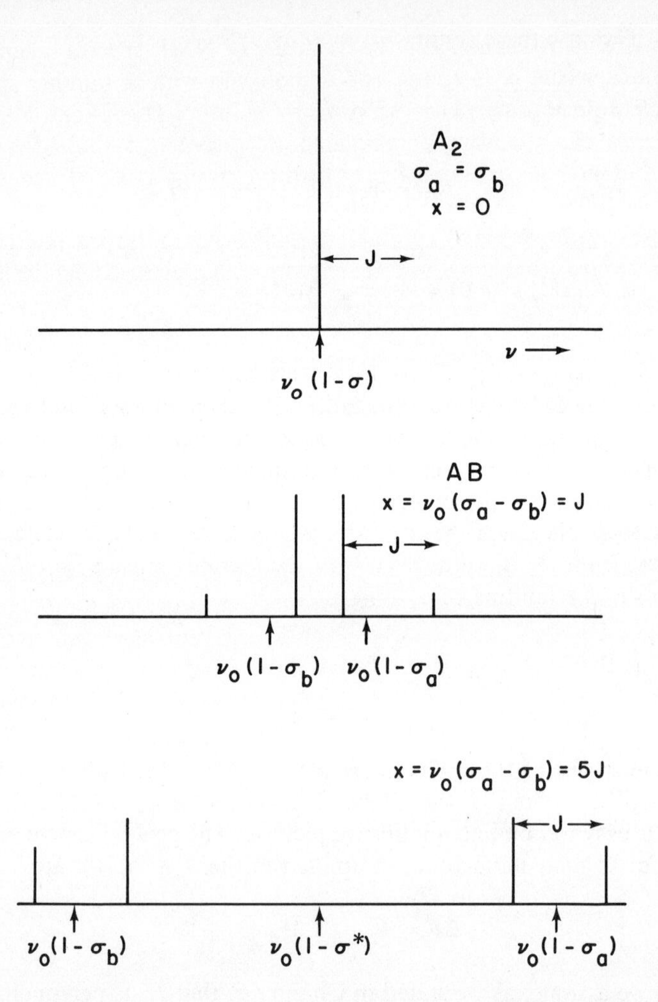

Figure 7.3. *AB spectrum for spin* $= \frac{1}{2}$. *In this drawing J and* σ^* *are held constant and* $\nu_0(\sigma_a - \sigma_b)$ *is varied. It can be seen that even though the intensities are far from equal when* $\nu_0(\sigma_a - \sigma_b)$ *is close to J, the splittings approach J very rapidly. The less intense pair of lines corresponds to the dotted transitions in Fig. 7.2.*

have the A_2 case and only one line located a $\nu_0(1 - \sigma)$. When $x^2 \gg J^2$ we have the AX case and there are two sets of equal doublets, one for each nucleus, split apart by J. When $x^2 \approx J^2$ the four possible transitions form a symmetrical pattern with two stronger and two weaker lines, as indicated in Fig. 7.2. A close examination of Fig. 7.3 shows that the splittings approach the AX limit much faster than do either the intensities or even the centers of each pair.

There are many examples in the literature of NMR spectra taken under conditions where $x \approx J$, but for a molecule with a number of nonequivalent protons there are a number of spin–spin couplings, and exact treatments can get very complicated. The chemical shifts for nuclei other than protons are quite large and this tends to make ^{19}F spectra, for example, follow the AX case. Since the inorganic literature is rich with examples of simple NMR spectra, we shall make the approximation that all our spectra are taken under sufficiently high magnetic fields to achieve uncoupling through the Paschen–Back effect.

PASCHEN–BACK HAMILTONIAN

At high magnetic fields Eq. (7.3) can be simplified so that the spectra of uncoupled nuclei can be derived. If two nuclei have quite different precessional frequencies, all the terms in the spin–spin interaction can be neglected except for the Z-axis one. The reasons for this are clear in the vector model because the X and Y products average to zero. If the spin–spin interaction is simplified so that only the Z-axis terms are used, this approximation gives what is called the first-order Hamiltonian. The neglected X and Y products can be added later in a perturbation treatment and they only contribute in higher than first order in a power-series expansion in J. For equivalent nuclei, however, the Paschen–Back effect never yields an uncoupling, for their precessional frequencies are identical. For this reason the first-order Hamiltonian should retain the full dot product for spin–spin coupling between equivalent nuclei.

If we neglect the X and Y terms between nonequivalent nuclei,

$$h^{-1}\mathsf{H}(\text{first order}) = -\sum_i \nu_{0i}(1 - \sigma_i)\mathbf{I}_{Zi} + \sum_{i<j} J_{ij}\mathbf{I}_{Zi}\mathbf{I}_{Zj} + \sum_{i<k} J_{ik}\mathbf{I}_i \cdot \mathbf{I}_k \qquad (7.9)$$

where i includes all nuclei, j includes only those which are not equivalent to i, and k includes those which are equivalent to i. We shall also assume that equivalent nuclei are completely equivalent in that they all have J_{ij} values equal to other equivalent nuclei. In some molecular geometries like 1,1-difluoroethylene this is not true, for there are two J_{FH} coupling constants, so that the hydrogens are not completely equivalent.

If now in Eq. (7.9) we first sum over equivalent nuclei, we can form sets of equivalent nuclei called A, B, and C with

$$\mathbf{F}_a = \sum_i^A \mathbf{I}_i$$

$$\mathbf{F}_b = \sum_i^B \mathbf{I}_i, \text{ etc.} \qquad (7.10)$$

We can then express Eq. (7.9) as

$$\begin{aligned} h^{-1}\mathsf{H}(\text{first order}) = & -\nu_{0a}(1-\sigma_a)\mathsf{F}_{Za} \\ & -\nu_{0b}(1-\sigma_b)\mathsf{F}_{Zb} - \cdots, \text{etc.} \\ & + J_{ab}\mathsf{F}_{Za}\mathsf{F}_{Zb} + J_{bc}\mathsf{F}_{Zb}\mathsf{F}_{Zc} + \cdots, \text{etc.} \\ & + \tfrac{1}{2}J_{aa}[\mathbf{F}_a^2 - 2n_aI_a(I_a+1)] \\ & + \tfrac{1}{2}J_{bb}[\mathbf{F}_b^2 - 2n_bI_b(I_b+1)] + \cdots, \text{etc.} \end{aligned} \tag{7.11}$$

where n_a are the number of equivalent nuclei with spin I_a, etc. In the representation where $\mathbf{F}_a^2$, F_{Za}, etc., are quantized, then

$$\begin{aligned} h^{-1}\epsilon(\text{first order}) = & -\nu_{0a}(1-\sigma_a)M_{F_a} \\ & -\nu_{0b}(1-\sigma_b)M_{F_b} - \cdots, \text{etc.} \\ & + J_{ab}M_{F_a}M_{F_b} + J_{bc}M_{F_b}M_{F_c} + \cdots, \text{etc.} \\ & + \tfrac{1}{2}J_{aa}[F_a(F_a+1) - 2n_aI_a(I_a+1)] \\ & + \tfrac{1}{2}J_{bb}[F_b(F_b+1) - 2n_bI_b(I_b+1)] \\ & + \cdots, \text{etc.} \end{aligned} \tag{7.12}$$

It can be shown that no transitions are allowed which change the F values, and for the NMR spectrum of the nuclei in set A the selection rule is

$$\Delta M_{F_a} = \pm 1, \qquad \Delta M_{F_b} = 0, \text{etc.}, \tag{7.13}$$

so that the spectrum of the nuclei in set A has a frequency given by

$$\begin{aligned} \nu_a &= \nu_{0a}(1-\sigma_a) - J_{ab}M_{F_b} \\ & \quad - J_{ac}M_{F_c} - \cdots, \text{etc.} \\ &= \nu_{0a}(1-\sigma_a) - \sum_R J_{aR}M_{FR}. \end{aligned} \tag{7.14a}$$

We can also go back to Eq. (7.10), where we can see that

$$M_{F_b} = \sum_i^B M_i, \text{etc.}, \tag{7.14b}$$

so that

$$\nu_a = \nu_{0a}(1-\sigma_a) - J_{ab}\sum_i^B M_i - J_{ac}\sum_i^C M_i - \cdots, \text{etc.} \tag{7.14c}$$

The spectra predicted by Eq. (7.14*a*) or (7.14*c*) are quite simple, but before these spectra are outlined it seems best to use the vector model to clarify this derivation. A vector-model diagram for Eq. (7.11) is shown in Fig. 7.4. In this figure we assume two sets of equivalent nuclei: set

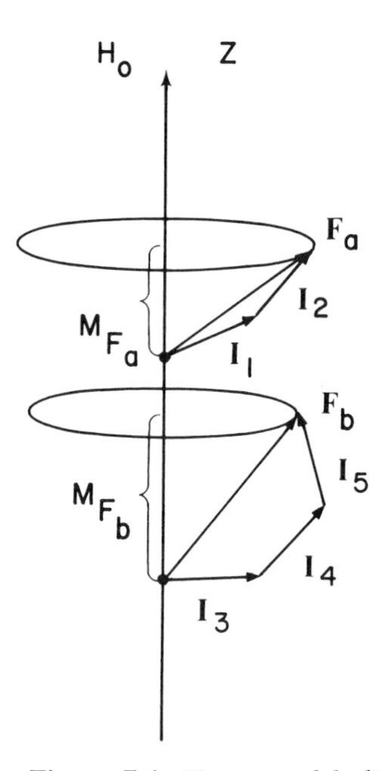

Figure 7.4. *Vector-model diagram for A_2X_3 spin–spin coupling. In the NMR transitions for spins A the values of* $\mathbf{F}_a^2$, $\mathbf{F}_b^2$, *and* M_{F_b} *do not change but* M_{F_a} *changes by* 1 *unit.*

A with two nuclei and set B with three. Since equivalent nuclei are always coupled, they precess together to form the vectors $\mathbf{F}_a$ and $\mathbf{F}_b$. In NMR transitions for the nuclei in set A, $\mathbf{F}_a$ is reoriented so that M_{F_a} changes by ± 1. In this reorientation the angle between the vectors within the set A remains unchanged, so that the energy term involving J_{aa} does not change. As a result the only terms in the energy for the NMR transitions of the set A are its own Zeeman term and the term containing J_{ab}. The only remaining question is how many possible values are there for $J_{ab}M_{F_b}$?

For each F value there are $2F + 1$ values of M_F, which range from $-F$ to $+F$ in value. However, for three spins with $I = \frac{1}{2}$, for example, there are several possible F values. It is best to consider first the two-spin problem, then the three-spin, etc. This is made easier by using Eq. (7.10) in steps, so that

$$\mathbf{I}_1 + \mathbf{I}_2 = \mathbf{F}_2$$
$$\mathbf{F}_2 + \mathbf{I}_3 = \mathbf{F}_3, \text{ etc.} \tag{7.15}$$

For two spins of $\frac{1}{2}$ the vector addition of angular momenta shows that $F_2 = 0$ or 1. The resulting M_F values for this addition process are shown in Table 7.2.

Table 7.2. *Number of Possible M_F Values for Spin $= \frac{1}{2}$ Splitting Diagram*

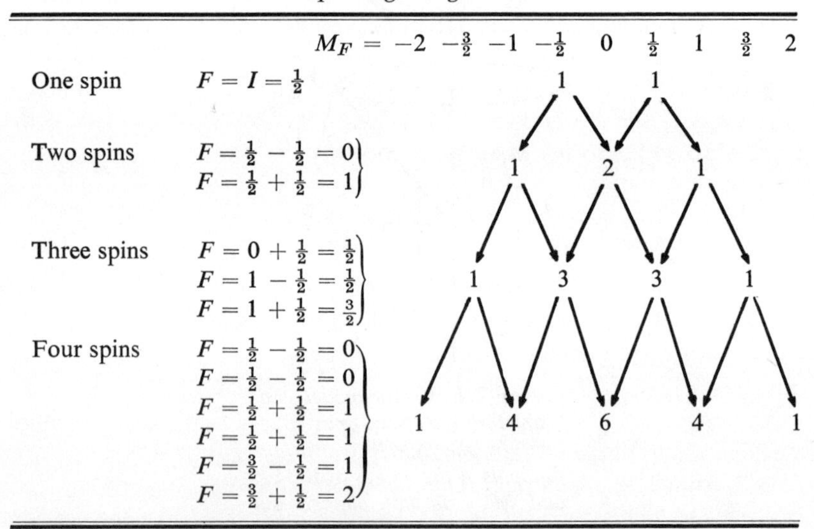

		$M_F = -2$	$-\frac{3}{2}$	-1	$-\frac{1}{2}$	0	$\frac{1}{2}$	1	$\frac{3}{2}$	2
One spin	$F = I = \frac{1}{2}$				1		1			
Two spins	$F = \frac{1}{2} - \frac{1}{2} = 0$ $F = \frac{1}{2} + \frac{1}{2} = 1$			1		2		1		
Three spins	$F = 0 + \frac{1}{2} = \frac{1}{2}$ $F = 1 - \frac{1}{2} = \frac{1}{2}$ $F = 1 + \frac{1}{2} = \frac{3}{2}$		1		3		3		1	
Four spins	$F = \frac{1}{2} - \frac{1}{2} = 0$ $F = \frac{1}{2} - \frac{1}{2} = 0$ $F = \frac{1}{2} + \frac{1}{2} = 1$ $F = \frac{1}{2} + \frac{1}{2} = 1$ $F = \frac{3}{2} - \frac{1}{2} = 1$ $F = \frac{3}{2} + \frac{1}{2} = 2$	1		4		6		4		1

It can be seen in Table 7.2 that three spin $= \frac{1}{2}$ states do yield two possible states with $F = \frac{1}{2}$. These together with the $F = \frac{3}{2}$ state give three possible states with $M_F = \pm\frac{1}{2}$. The four-spin case is more complicated, but quick examination shows that the number of possible ways to get a given M_F value are given by the binomial coefficients. This is only true for spin $= \frac{1}{2}$, however. The arrows in Table 7.2 illustrate how Eq. (7.15) for spin $= \frac{1}{2}$ simply splits an M_F for one less spin into two M_F values. This is also an illustration of Eq. (7.14*b*), where each additional spin $= \frac{1}{2}$ forms two new M_F values. The arrows in Table 7.2 are also a common way to generate the binomial coefficients.

Table 7.2 also forms a splitting diagram. According to Eq. (7.14*a*) the B nuclei split the NMR resonance of set A according to the M_F values of

B. According to Table 7.2, if, as assumed in Fig. 7.4, the set B contains three equivalent nuclei with spin $= \frac{1}{2}$, the A resonance is split into four equally spaced lines with intensity 1:3:3:1. According to the first-order Hamiltonian the splitting constant is J_{ab}. On the other hand, the NMR of the B nuclei would be split by the A nuclei into a 1:2:1 triplet with the same splitting constant J_{ab}. We can see that if sets A and B form an A_2X_3 case as previously discussed, very simple splittings are observed.

7.2. *Examples of High-Resolution Spectra*

In NMR instruments either the frequency is held essentially fixed and the magnetic field H_0 is swept, or the field is held essentially fixed and the frequency is swept. There are advantages to both methods of operation, but we will always assume that all spectra are taken with a field sweep. Absolute field measurements are rarely made, and the most common practice is to use a reference substance and to measure field shifts from it. A reduced field shift δ can be defined as

$$\delta(\text{ppm}) = \pm \frac{H_0(\text{sample}) - H_0(\text{ref})}{H_0(\text{ref})} \times 10^6. \tag{7.16}$$

The plus and minus sign expresses the fact that both conventions are used. For a fixed frequency and a single isotope Eq. (7.16) assumes that absorption will occur when $H(\text{sample}) = H(\text{ref})$. Therefore, with the positive sign in Eq. (7.16), δ will be positive if the sample nuclei are more shielded than are the reference nuclei.

For protons the common reference is tetramethylsilane (TMS). Its protons are highly shielded, so that to make δ positive the negative sign is usually used in Eq. (7.16) for protons vs. TMS. In fact, most organic molecules will then have δ values that vary from 1.0 (high shielding) to 9.0 (low shielding). In order to have a reduced field shift that increases with shielding, many workers prefer to use τ where

$$\tau(\text{protons}) = 10 - \delta(\text{TMS}). \tag{7.17}$$

In this case the common organic protons would then vary from $\tau = 9.0$ (high shielding) to $\tau = 1.0$ (low shielding). An unshielded proton, for example, has been measured to be -21 on the τ scale or $+31$ on the δ scale vs. TMS.

Liquids can be run without a solvent (neat) or in a solvent that does not give an interfering NMR spectrum. Solvents like CS_2 or CCl_4

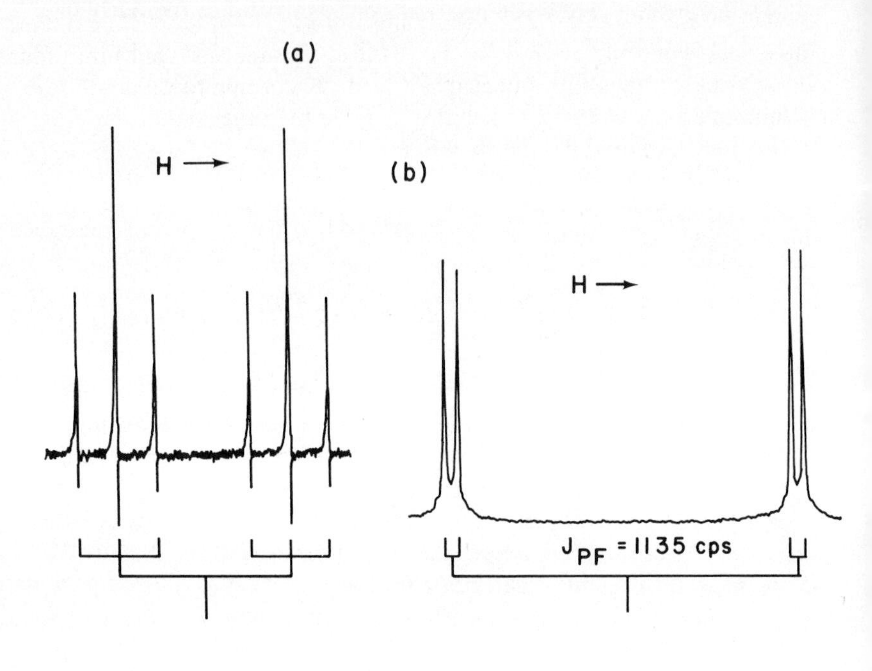

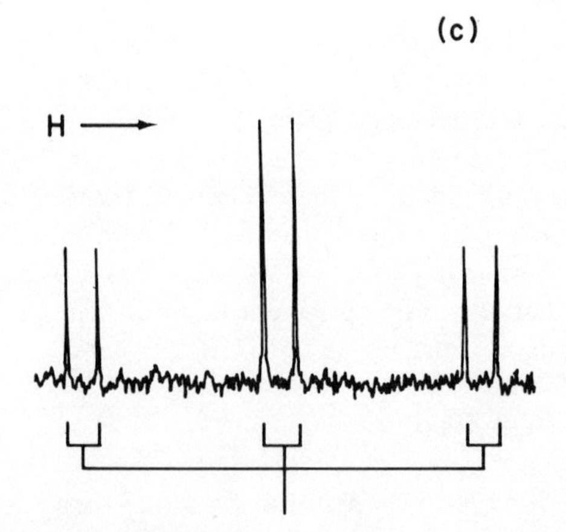

Figure 7.5. *NMR spectra of liquid* PHF_2 *at* $-20°C$. (*a*) ^1H *spectrum at* 60 *MHz*. (b) ^{19}F *spectrum at* 94.1 *MHz*. (*c*) ^{31}P *spectrum at* 40.4 *MHz*. (*Courtesy R. Rudolph and R. Parry.*)

contain no protons, but when necessary one can also use the fairly inexpensive deuterated solvents, such as D_2O, $CDCl_3$, or C_6D_6. It is observed that some chemical shifts depend upon the solvent, so that it is best to standardize on a single solvent for classes of molecules.

For ^{19}F, ^{13}C, etc., there have been a variety of reference compounds used. This is partially because these nuclei have large chemical shifts, so that it is not as convenient as it is for protons to standardize on a single reference compound for each nucleus. However, a common reference for ^{19}F is trifluoracetic acid and a common ^{13}C reference is benzene or enriched CH_3I. All these liquids contain protons and the magnetic field or frequency can be "locked on" these protons while taking ^{19}F or ^{13}C spectra.

PHF_2

In the molecule PHF_2 high-resolution spectra can be taken for all three kinds of nuclei since all have spin $= \frac{1}{2}$. In Fig. 7.5*a* the proton spectrum is shown. This should be an excellent first-order spectrum since different kinds of nuclei are involved. It can be seen that J(PH) splits the proton spectrum into doublets and J(FH) further splits it into 1:2:1 triplets, as Eq. (7.14) predicts. This would still be an excellent first-order spectrum even if $J(\mathrm{PH}) = J(\mathrm{FH})$, since the important fact is the large difference between the Larmor frequencies of ^1H, ^{19}F, and ^{31}P compared with the J values.

In Fig. 7.5*b* we show the ^{19}F spectrum. Since the two fluorines are equivalent, they form an A_2 system and J(FF) does not appear in the spectrum, so that the ^{19}F spectrum is split into double doublets. The smaller splitting equals J(FH) and is the same, within experiment error, in both Figs. 7.5*a* and 7.5*b*. The large J(PF) is, of course, the dominant splitting in Fig. 7.5*b*. The ^{31}P spectrum gives no new information, but it is also shown in Fig. 7.5*c*. The J(PF) now forms a 1:2:1 triplet with essentially the same splitting observed for ^{19}F. The smaller doublet splitting is equal to J(PH) seen first in the proton spectrum.

The signs of the J values are not important in the first-order Hamiltonian, but their relative signs can be important in the complete Hamiltonian. Simple bonding theory indicates, as shown in Fig. 7.1, that the J value between most directly bonded nuclei should be positive in sign. This assumes that the nuclear moments are positive. Of the first-row elements fluorine is an exception in yielding negative J values. On the basis of rather complete NMR measurements it appears for PHF_2 that J(PH) and J(PF) have the opposite sign but that J(PH) and J(HF) have the same sign. It is also very likely that J(PH) is positive, so that the signs of all these J values are established. However, since the signs of the

J values have little importance, in most instances we will not be concerned with them.

$CF_3—S—CFH—CF_2H$

The complicated molecule $CF_3—S—CFH—CF_2H$ is a good example of a first-order proton spectrum. The J value between the two protons is sufficiently small compared to the chemical shift at 60 MHz so that Eq. (7.14) can be used. This molecule is a good example of how J values drop off with the number of bonds, but for ^{19}F this drop-off can also be rather slow.

A glance at Fig. 7.6 shows that there are two symmetrical splitting patterns. The middle lines are part of a 1:2:1 triplet which is split into

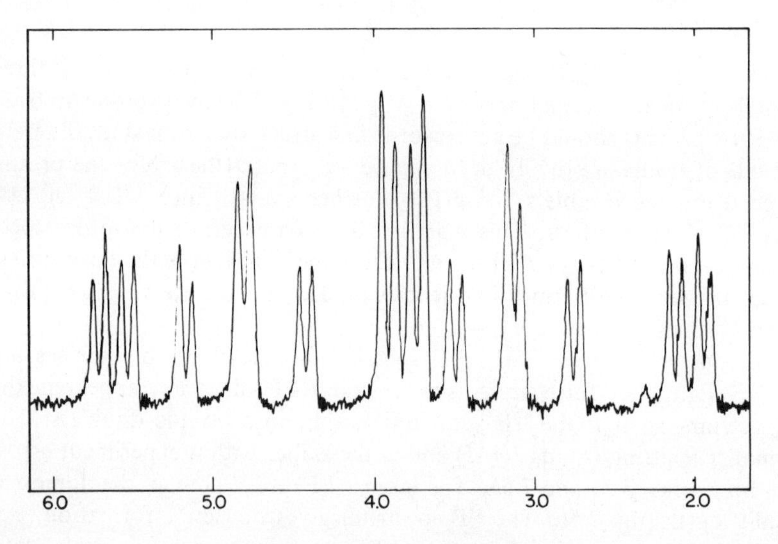

Figure 7.6. *Proton spectrum of* $CF_3—S—CFH—CF_2H$ *at 60 MHz. The spin-spin coupling constants determined from both the* ^1H *and* ^{19}F *spectrum are* $J_{ab} = 3.9, J_{ac} = 1.5, J_{bc} = 19.5, J_{db} = 48.7, J_{dc} = 11.0, J_{ce} = 54.5. J_{be} = 5.0,$ *and* $J_{de} = 2.5$, *all in Hz. The designations follow:*

$$\overset{a}{CF_3}—S—\overset{b}{C}\overset{d}{F}H—\overset{c}{C}\overset{e}{F_2}H.$$

(Courtesy L. Petrakis and C. Sederholm.)

what looks like a 1:1:1:1 quartet. The second pattern is a doublet split into a 1:2:1 triplet where each line is also double. From the molecular structure, and the assumption of larger J values for geminal atoms, the first set is assigned to the end proton and the second to the middle proton. A 1:1:1:1 quartet is not possible with spin $= \frac{1}{2}$, so that it must be due to double doublets with one splitting about twice the other.

We shall not complete this assignment but leave it to the reader. The actual J values are given in Fig. 7.6, where J in hertz can be obtained by multiplying the splittings in parts per million by 60 since this is a 60-MHz spectrum. One notes that the only J values that can be observed for the CF_3 group are in the ^{19}F spectrum. Some of the intensity variations in Fig. 7.6 must be due to the terms in the second-order Hamiltonian.

ClF_3

The ^{19}F NMR spectrum of the molecule ClF_3 at low temperatures shows two distinct multiplets. At first glance one multiplet is a doublet and the other is close to a 1:2:1 triplet. Both the common chlorine isotopes have large nuclear quadrupole couplings and their spin–spin interactions are averaged to zero in liquids (see Fig. 6.7). The observed spin–spin splittings indicate two kinds of fluorine atoms. Several structural methods show that ClF_3 is close to a T-shaped molecule with the chlorine at the center and a fluorine on each end.

Figure 7.7 shows the 40-MHz spectrum of liquid ClF_3. The additional splittings must be due to the second-order Hamiltonian, for they almost

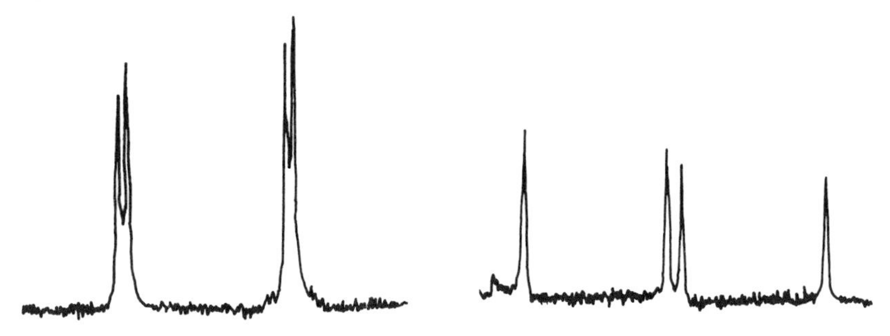

Figure 7.7. ^{19}F *spectrum of liquid* ClF_3 *at* $-63°C$ *and* 40 *MHz. Two portions are shown with different sweep rates. (Courtesy L. Alexakos and C. Cornwell.)*

disappear at 60 MHz. At 40 MHz it is clearly an AB_2 spectrum, whereas above 60 MHz it is essentially AX_2. The spectrum of gaseous ClF_3 can also be observed, and at high pressure, where the rotational correlation time is short, essentially the same spectrum can be observed.

PF_5 AND $(CH_3)HPF_3$

In the early days of high-resolution NMR H. S. Gutowsky and C. J. Hoffman (1951) observed the ^{19}F spectrum of PF_5 to be a doublet but with unequal intensities. This seemed reasonable since spin–spin splitting was not known at that time and PF_5 has a trigonal bipyramidal structure with three equatorial fluorines and two axial fluorines. In subsequent work H. S. Gutowsky, D. W. McCall, and C. P. Slichter (1953) showed that the two ^{19}F signals from PF_5 really had equal intensity and that the splitting was also the same at higher magnetic fields. They clearly showed that this splitting was due to spin–spin coupling with the ^{31}P.

The problem remains as to why only one type of fluorine seemed to be present. The answer is that there is exchange between the equatorial and

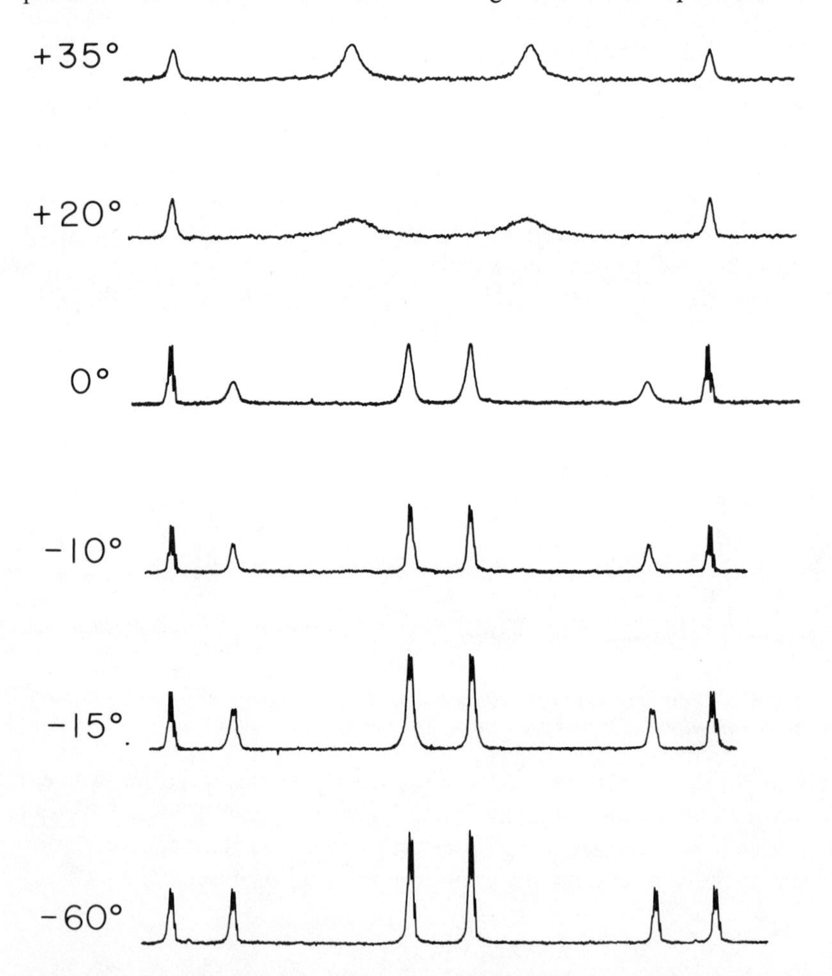

Figure 7.8. *Portion of the proton spectrum of* $(CH_3)HPF_3$. *This is the lower-field component of the* ^{31}P *doublet for the proton bonded to the phosphorus.* [*Courtesy R. A. Goodrich and P. M. Treichel; see Inorg. Chem.*, **7**, 694 (1968).]

axial positions. If the correlation time associated with this exchange is much shorter than $\delta\omega$ (the difference between the two Larmor frequencies), only a single ^{19}F species can be detected in the NMR. This condition can be met in many chemical systems, and so it is not uncommon to lose chemical shifts or spin–spin splittings because of exchange. Since the ^{31}P spin–spin splitting is not lost in PF_5, one knows that it must be an intramolecular exchange and not an intermolecular exchange. Fast exchange between molecules will also "average out" the spin–spin splittings since the average I_Z value is essentially equal to zero.

Figure 7.8 shows part of the ^{1}H spectrum of $(CH_3)HPF_3$ over a range of temperatures. This is only one-half of a ^{31}P doublet, and only the proton that is bonded to the phosphorus is being observed. At low temperatures one sees a 1:2:1 triplet split into a doublet. The very close fine structure is due to a small spin–spin coupling with the CH_3 protons. There are several possible isomers for this molecule, but investigations have shown that the most stable species has only fluorines in the two axial positions, as shown in Fig. 7.9*a*. This agrees with the low-temperature spectrum in Fig. 7.8, where we would only have to assume that $J(H, F_{ax}) > J(H, F_{eq})$ to explain the doublet and triplet structures.

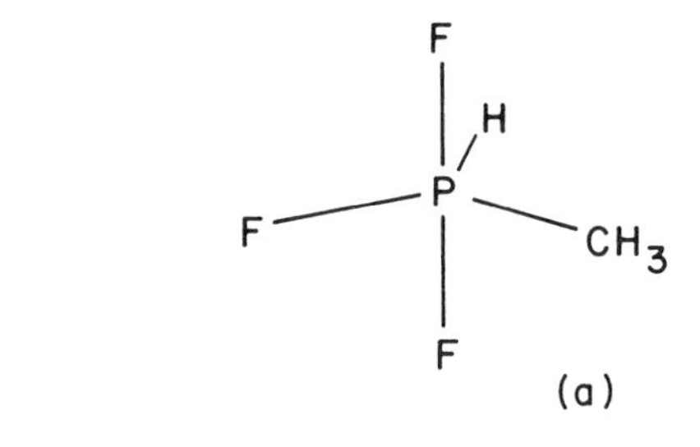

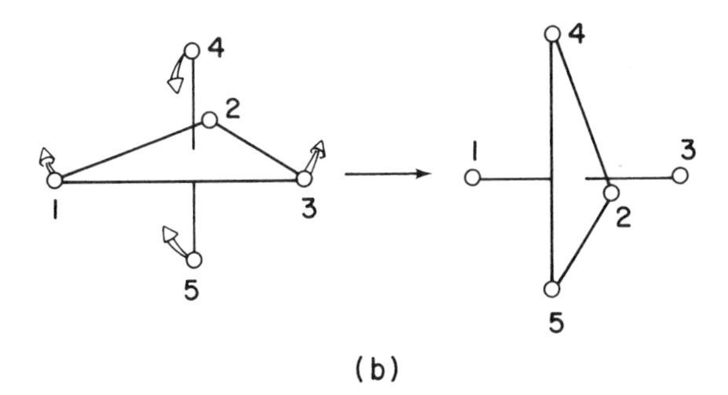

Figure 7.9. *Structure and dynamics of* $(CH_3)HPF_3$. (*a*) *The most stable isomer.* (*b*) *The dynamics of pseudorotation about an equatorial atom.* (*After R. S. Berry.*)

At high temperatures the spin–spin splitting pattern approaches a 1:3:3:1 quartet. This pattern would be caused by three equivalent fluorine atoms. An exchange between equatorial and axial positions for the fluorines would make all three equivalent and explain the high-temperature spectrum. Like the related PF_5 this must be an intramolecular exchange. A mechanism for this exchange is shown in Fig. 7.9*b*. In this diagram a pseudorotation takes place about one of the equatorial atoms which exchanges an axial and an equatorial pair. Since the most stable isomer only has fluorines in the axial positions, the pseudorotation must continue about another equatorial atom so that two fluorines are returned to the axial position and are exchanged in the process.

H_2O, H_3O^+, AND NH_3

When acids are dissolved in water only a single proton NMR spectrum is observed. Its shielding depends upon the relative amounts of acid and H_2O, and it has no spin–spin splitting. This is a clear case of intermolecular exchange, with H_3O^+ or a similar species as the intermediate in the exchange.

When ordinary 99.99 per cent pure liquid ammonia is observed in a high-resolution proton NMR apparatus, a single, fairly sharp line is observed. Since ^{14}N has $I = 1$, one might expect a spin–spin triplet, but one might also think that the nuclear quadrupole coupling with the ^{14}N nucleus averages the spin–spin splitting to zero. This is a reasonable guess since spin–spin splitting is not often observed from ^{14}N, but if the ammonia is very carefully purified, then a nice spin–spin triplet is observed.

Figure 7.10 shows the proton spectrum of liquid ammonia containing a large amount of water and some hydroxide ions. Separate spectra are observed for the NH_3 and for the $H_2O + OH^-$. The water resonance is broadened by exchange with the OH^- but the ^{14}N splitting is observed for the NH_3. The purpose of the OH^- in this experiment was to reduce the NH_4^+ to a very low level, for it is this species that produces the rapid exchange in ordinary liquid ammonia. The reaction is

$$NH_4^+ + NH_3 \rightleftharpoons NH_3 + NH_4^+.$$

Near $\delta = 0$ and $\delta = 1$ (see Fig. 7.10) small sharp signals can be observed. This is a doublet that is due to $^{15}NH_3$ in natural abundance. The fact that these lines are much sharper than those due to $^{14}NH_3$ shows that the $^{14}NH_3$ signals are broadened somewhat by quadrupole relaxation. This is also shown by the fact that the center of the ^{14}N triplet is narrower than the outer two. The theory of quadrupole relaxation in liquids shows

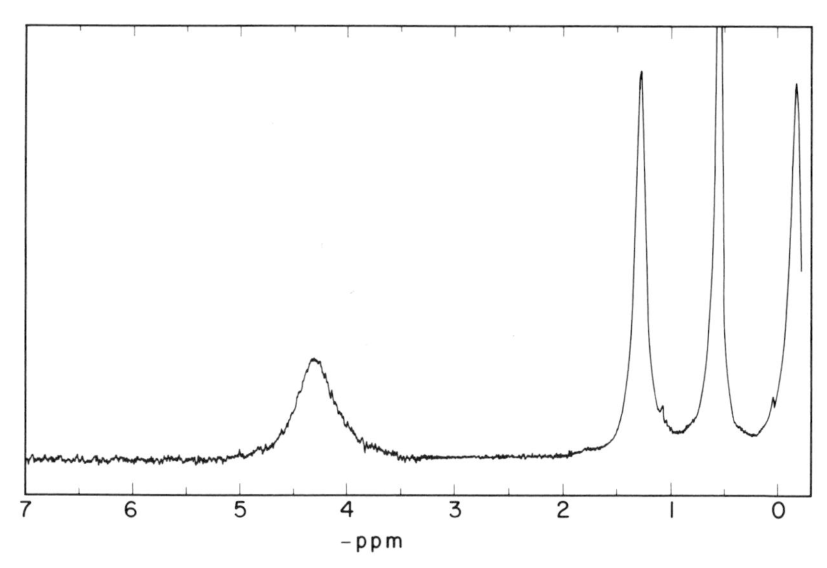

Figure 7.10. *Proton spectrum of liquid ammonia containing water and* OH^-. *The solution was saturated in* NaOH *and* 0.86 *M in* H_2O. [*Courtesy T. Birchall and W. L. Jolly; see J. Amer. Chem. Soc.*, **87**, 3007 (1965).]

that the central line is broadened less than are the outer ones, because for very rapid relaxation the entire spin multiplet collapses into a sharp central line.

7.3. Broad-Line Spectra in Liquids

When $I \geq 1$, nuclear quadrupole relaxation in liquids will give broadened NMR spectra for these nuclei. However, there are a number of circumstances that can also give broadened lines. These circumstances can generally be summarized as those cases where the nucleus is exposed to a fluctuating magnetic field H^* but where the correlation time τ_c associated with this fluctuation is not short enough to give a very narrow line. Important chemical information can be obtained from NMR in these cases, since the line width is then a source of information about various fast processes in solution.

^{17}O IN WATER SOLUTIONS

A great deal of information about fast chemical reactions has been obtained from a study of ^{17}O NMR in water solutions. If a small amount of a paramagnetic ion is dissolved in water, the ^{17}O resonance is greatly broadened. This is because some of the water molecules coordinate around the paramagnetic ion and are exposed to a local magnetic field. Several processes can give fluctuations to this local field. One of them is chemical exchange with the bulk water molecules. A temperature study of the ^{17}O line width can establish the source of the fluctuations and determine the rate constants that are involved [see T. J. Swift and R. E. Connick, *J. Chem. Phys.*, **41**, 2553 (1964) and **37**, 307 (1962)].

If the exchange is slow, it should be possible to see the NMR spectrum of the ^{17}O directly coordinated to an ion in solution. By measuring the intensity of this ^{17}O signal it is possible to determine the number of water molecules coordinated around this ion. Fig. 7.11 shows the ^{17}O spectrum of an enriched water solution containing a small amount of Co^{+2}

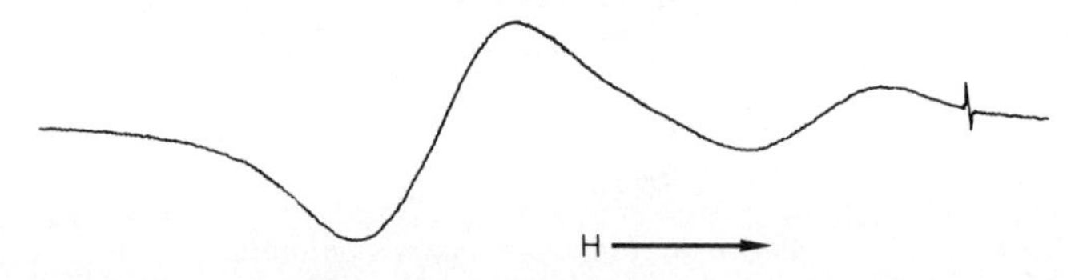

Figure 7.11. ^{17}O *resonance of water and hydrated* Al^{+3}. *The water was enriched to* 11.5 *per cent in* ^{17}O *and was* 1.5 *M in* Al^{+3} *and* 0.2 *M in* Co^{+2}. *On the side of the higher-field signal due to hydrated* Al^{+3} *is the much sharper signal of a small sample of pure* ^{17}O *enriched water. The signals presented are the first derivative of the absorption.* [*Courtesy R. E. Connick and D. N. Fiat, see J. Chem. Phys.*, **39**, 1349 (1963).]

and a large amount of Al^{+3}. The water bound to Co^{+2} exchanges very rapidly with the bulk water molecules and gives a single ^{17}O signal. The purpose of adding the Co^{+2} is to shift the resonance of the bulk water so that it is away from the signal from the water coordinated around the Al^{+3}.

The Al^{+3} exchanges its water very slowly and so its water has a separate resonance, as shown in Fig. 7.11. If one determines the relative intensities of the bulk water and the coordinated Al^{+3} signals, the coordination number of the Al^{+3} can be evaluated. The experimental results from a number of spectra showed that the Al^{+3} is best represented as $Al(H_2O)_6^{+3}$ in aqueous solution. This confirms our assumption of octahedral coordination for transition-metal ions in aqueous solution since Al^{+3} is close in size to many transition-metal ions.

^{35}Cl IN VCl_4

In a large number of cases, for nuclei with $I \geq 1$, nuclear quadrupole relaxation in liquids can give basic line widths of more than 1 gauss;

Table 7.3. *Correlation of* ^{35}Cl *Line Width and Quadrupole Coupling Constant for Liquids*[a]

Compound	Line width[b] (gauss)	Quadrupole coupling constant (MHz)
$TiCl_4$	2.2	12
$VOCl_3$	5.5	22
CrO_2Cl_2	8.3	32
$SiCl_4$	14.7	40

[a] Adapted from Abragam.
[b] Line width defined as $2/\gamma T_2$.

^{35}Cl is rather typical of this case. Table 7.3 shows some ^{35}Cl widths for a series of related molecules and how they are correlated to the ^{35}Cl quadrupole coupling constants. The quadrupole coupling constant is the product of the nuclear quadrupole moment and the gradient of the electric field at the nucleus in each compound. One can show that the gradient of the electric field at the nucleus is primarily determined by the distribution of the p electrons. A filled p shell is spherically symmetrical, and so the electric field gradient at a nucleus is related to the unfilled nature of the p orbitals.

In Table 7.3 one can see that larger quadrupole couplings lead to larger line widths. This should be true as long as the correlation time associated with tumbling is relatively constant. This would be expected in liquids with the same molecular size and viscosity such as is found in Table 7.3. A quantitative calculation similar to Eq. (6.25) shows that there is good agreement between the expected rotational correlation times and those necessary to explain the line widths in Table 7.3 (see Problem 7.9).

The ^{35}Cl resonance in VCl_4 is more complicated than it is in $TiCl_4$, since VCl_4 is paramagnetic. The chlorine nuclei are exposed to a large local magnetic field. If this field did not vary with time, it would simply give very large shifts to the ^{35}Cl resonance. But in VCl_4 there are fast fluctuations to this local field.

The major source of fluctuations to the local ^{35}Cl field is due to spin–lattice relaxation for the odd electron in VCl_4. Spin–lattice relaxation tends to average the S_Z value to zero because it corresponds to rapid changes in the values of M_S in the vector model. If S_Z averaged to exactly zero, the local ^{35}Cl field would only lead to a broadening of the ^{35}Cl resonance. However, S_Z does not average to exactly zero, so it also provides a shift in the resonance.

These arguments can be put on a quantitative basis with the use of the spin Hamiltonian. The necessary spin Hamiltonian for paramagnetic liquids is discussed in Chapter 8. One can show that the energy of a nucleus in a liquid paramagnetic is well represented by

$$\epsilon(\text{nucleus}) = AM_I M_S + \frac{\gamma H_0}{2\pi} M_I \tag{7.18}$$

where γ is the magnetogyric ratio for the nucleus, including the effects of chemical shift, and A is the Fermi contact interaction with the paramagnetic electron. If one used Eq. (7.18) as the basis of an NMR spectrum with the selection rules $\Delta M_I = \pm 1$, the first term would contribute a large zero-field splitting and give an unusual NMR pattern. For most systems one would not be able to measure such a spectrum, for the line widths would be so large that the lines could not be detected. This is because the first term in Eq. (7.18) is so large that any slow fluctuations associated with this term give a width close to the Larmor frequency.

In a paramagnetic molecule with an electronic T_1 such that $1/T_1 \gg \hbar A$ a narrowing of the resonance is observed. This narrowing is due to a rapid averaging of the two NMR lines predicted by Eq. (7.18). Spin–lattice relaxation forces the nucleus to see an average orientation of the electron between $M_S = -\frac{1}{2}$ and $M_S = +\frac{1}{2}$, so that

$$\langle\epsilon(\text{nucleus})\rangle = AM_I\langle M_S\rangle + \frac{\gamma H_0}{2\pi} M_I. \tag{7.19}$$

The first term in this equation could be neglected if the $\langle M_S\rangle$ were essentially equal to zero, but for a paramagnetic in a magnetic field $\langle M_S\rangle$ is far enough from zero to make the first term still significant. For $S = \frac{1}{2}$ one can use Eq. (3.2c) and obtain

$$\langle M_S\rangle = -\frac{g\mu_B H_0}{4kT}, \tag{7.20a}$$

where a general g is substituted for the g_0 in Eq. (3.2c). One now obtains

$$\langle\epsilon(\text{nucleus})\rangle = \left(\frac{\gamma}{2\pi} - \frac{g\mu_B A}{4kT}\right) H_0 M_I. \tag{7.20b}$$

The second term in Eq. (7.20b) is like a chemical shift except that it clearly depends upon temperature. Both the magnitude and the sign of A can be obtained from a temperature study of the shift predicted by Eq. (7.20b).

The results of NMR measurements on ^{35}Cl in pure liquid VCl_4 is shown in Fig. 7.12. The value of the slope for this curve gave $A/h = -1.4 \times 10^6$ Hz where g was found by low-temperature EPR to be 1.93.

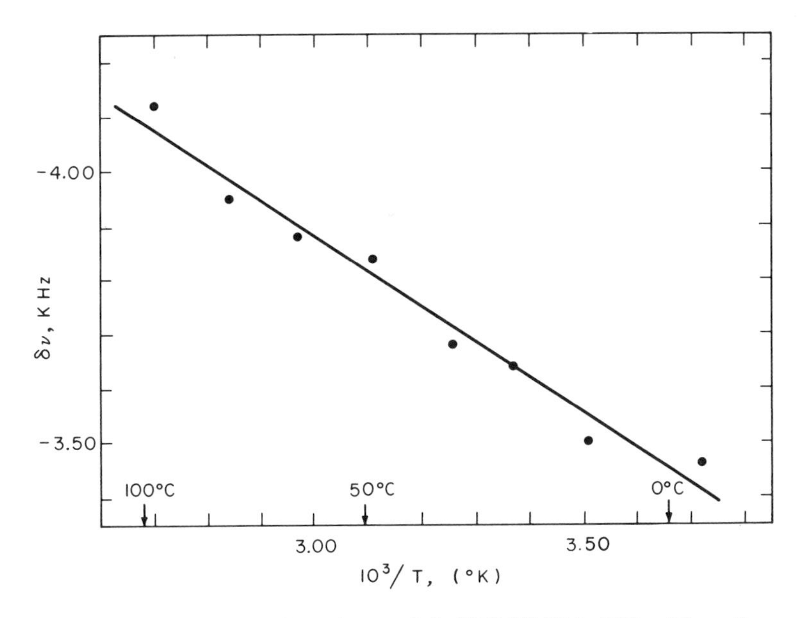

Figure 7.12. *Temperature dependence of the* ^{35}Cl *NMR in* VCl_4. *The reference was aqueous* Cl^- *and* H_0 *was* 14.09 *KG. From the slope of the line one obtains* $A/h = -1.4 \times 10^6$ *Hz.*

The line widths of this resonance could be used to also estimate the small electronic T_1 with an altered form of Eq. (6.25). In Chapter 8 we shall see that the small value for the electronic T_1, which makes it possible to see the NMR lines, also makes it impossible to see any direct EPR spectra at this temperature.

7.4. Broad-Line Spectra in Solids

In the solid state a nucleus experiences all the interactions that are present in liquids except that they are now anisotropic. Therefore, the chemical shift and spin–spin splitting depend upon the relative orientation of the magnetic field and the molecule. These anisotropic effects can be completely lost, however, in the large anisotropic dipole–dipole interactions between nuclei.

As Fig. 5.5 shows, each nucleus produces an anisotropic magnetic field due to its own magnetic dipole moment. When a nucleus is a few angstrom units away from another one, it will experience an anisotropic magnetic field which can be as high as 10 gauss. There is therefore a dipole–dipole interaction between two nuclei. The tumbling in liquids reduces this dipole–dipole interaction to zero since, as Fig. 5.5 shows, the average field around a dipole is zero. The averaging of the dipole field to zero is exact as long as the nuclei are uncoupled by the Paschen–Back effect and as long as the molecules in the liquid tumble rapidly. With little averaging of dipole–dipole interaction most nuclear resonances in solids are several gauss wide.

The shape of a nuclear resonance in a solid depends upon the distribution of the nuclei which surround each other. The overall resonance line more often has a Gaussian line shape, since the distribution of nuclei in solids are somewhat statistical. In some cases the magnetic nuclei are mainly in close pairs or triplets, and in these cases separate broad lines can be resolved from each other. The protons in single crystals of $CaSO_4 \cdot 2H_2O$ are found to form separate peaks about 10 gauss apart. Only the two protons in each water molecule have magnetic moments. In such favorable cases the proton–proton distance can be determined quite accurately for the water molecules in this simple hydrate.

In many solids, however, only a single broad resonance is observed. J. H. Van Vleck (1948) showed that the second moment of an NMR line in a solid can be calculated fairly easily for solid samples. The second moment $(\Delta H^2)_{\text{av}}$ is the average square deviation of the absorption signal from its center. It is equal to the mean-square deviation of the field weighted by the line-shape function. For a perfect Gaussian line shape $(\Delta H^2)_{\text{av}}$ is equal to the square of its standard deviation.

Van Vleck's result for a crystal with only one kind of magnetic nuclei and with n per unit cell of the crystal is

$$(\Delta H^2)_{\text{av}} = \frac{3}{4n} g_I^2 \mu_N^2 I(I+1) \sum_{ij} \frac{(1 - 3\cos^2\theta_{ij})^2}{r_{ij}^6} \tag{7.21a}$$

where θ_{ij} is the angle that the radius vector $\mathbf{r}_{ij}$ between two nuclei i and j makes with the magnetic field. In a cubic or other lattice in which all n nuclei are exactly equivalent, the sum over i in Eq. (7.21a) gives n identical terms, so that

$$(\Delta H^2)_{\text{av}} = \frac{3}{4} g_I^2 \mu_N^2 I(I+1) \sum_{j} \frac{(1 - 3\cos^2\theta_{ij})^2}{r_{ij}^6} \tag{7.21b}$$

Almost as much information can be obtained from a spectrum of a powdered or polycrystalline sample, and for such a system with more than one kind of nuclei

$$(\Delta H^2)_{\text{av}} = \frac{3}{5n} g_I^2 \mu_N^2 I(I+1) \sum_{ij} r_{ij}^{-6} + \frac{4}{15n} g_{I'}^2 \mu_N^2 I'(I'+1) \sum_{ik} r_{ik}^{-6} \qquad (7.22)$$

where the second term is for a second kind of nuclei interacting with those nuclei whose NMR is being observed. Any molecular motion that is fast compared to the difference between the Larmor frequencies of nuclei is important in considering the averaging in going from Eqs. (7.21*a*) to (7.22). In obtaining Eq. (7.22) we have averaged $(1 - 3\cos^2\theta_{ij})^2$ over all possible values, with the assumption that it was a simple average in three-dimensional space. If in the solid state there is large-amplitude molecular motion which averages $(1 - 3\cos^2\theta_{ij})^2$ even in a single crystal, the simple space average given by Eq. (7.22) is incorrect. For if a molecular unit in the solid rotated as in a liquid, some intramolecular contributions to $(\Delta H^2)_{\text{av}}$ are reduced to zero before any polycrystalline averages are taken. Molecular motion in a solid will reduce the $(\Delta H^2)_{\text{av}}$ value below that predicted by Eq. (7.22). There are many examples in the literature of the detection of large-amplitude motions in solids by NMR.

SF_6 IN WATER CLATHRATE

Water forms a cage or clathrate structure around dissolved gas molecules at low temperature. These hydrates are stable solid species and they can have both definite structures and molecular formulas. If, for example, SF_6 is bubbled into water at 0°C and several atmospheres of pressure, a clathrate with the formula $SF_6 \cdot 17H_2O$ precipitates. It is known, however, that the SF_6 molecules fill sites in the hydrate lattice that are relatively large, and one would expect that the SF_6 molecules could possibly have both some rotational and translational freedom.

In Fig. 7.13 we show the ^{19}F NMR line shape, in derivative form, for SF_6 in its clathrate. When this fairly broad line is integrated, the resultant absorption signal is between Gaussian and Lorentzian, but the value of $(\Delta H^2)_{\text{av}}$ calculated from the absorption signal under these conditions was 1.00 (gauss)^2. In Fig. 7.14 a plot is given of the second moment values obtained for SF_6 both in the ordinary clathrate and in the clathrate formed with deuterated water.

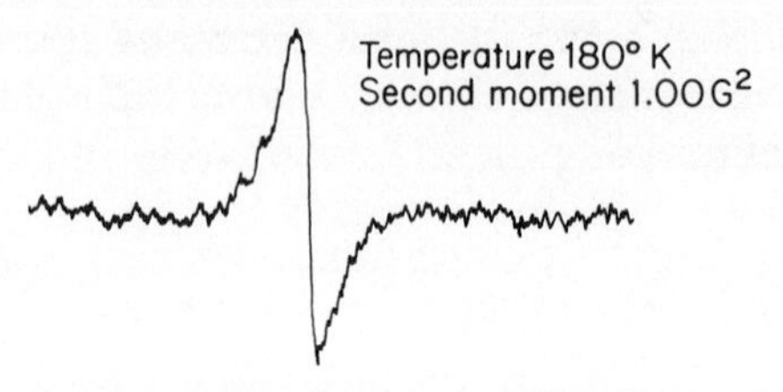

Figure 7.13. ^{19}F *spectrum of* SF_6 *in deuterated clathrate. The signal shown corresponds to the first derivative of the absorption.* [*From C. A. McDowell and P. Raghunathan, Molec. Phys.*, **13,** 331 (1967).]

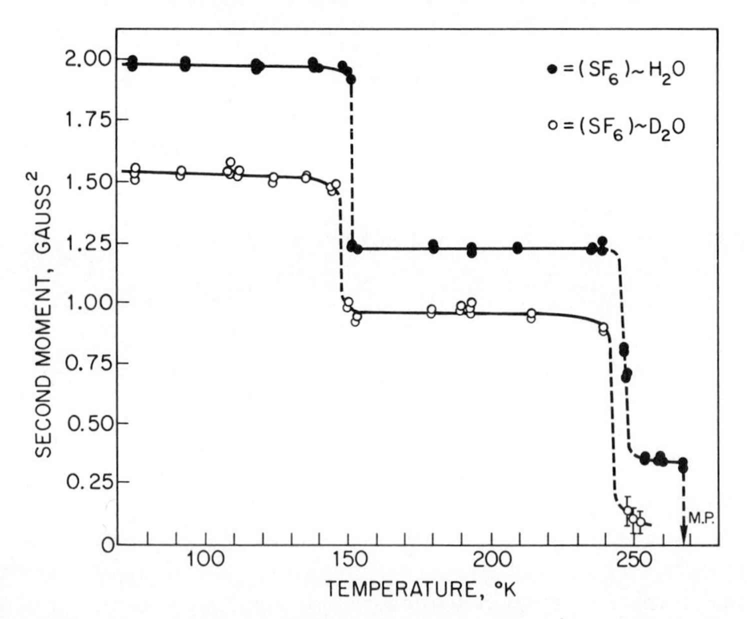

Figure 7.14. *Second-moment values for* SF_6 *clathrate over a range of temperature.* (*From McDowell and Raghunathan.*)

It is clear from Fig. 7.14 that there are both intra- and intermolecular contributions to $(\Delta H^2)_{av}$. The magnetic contributions predicted by Eq. (7.22) for ^{1}H are 16 times those predicted for ^{2}H, and so to a good approximation the values for SF_6 in deuterated clathrate must be entirely due to the SF_6 molecules themselves. If one also uses Eq. (7.22) for a single rigid SF_6 molecule with an octahedral structure and an S—F bond distance of 1.58 Å, it predicts a $(\Delta H^2)_{av}$ value of nearly 13 (gauss)^2. The sulfur nuclei can be neglected in this calculation since all common sulfur isotopes have $I = 0$.

One can conclude from the results in Fig. 7.14 that the SF_6 must rotate in some way even at 80°K. In addition, one can conclude that either additional rotational or translational degrees of freedom become possible at higher temperatures. There are several possible ways that octahedral SF_6 could have a restricted rotation. Any of these ways would be expected to give a $(\Delta H^2)_{av}$ value between 1 and 3 (gauss)^2, as is observed between 80 and 240°K. Above 240°K we would predict that there is nearly a completely free rotation of the SF_6 molecules in the clathrate.

NARROWING TECHNIQUES

It is possible to remove most of the effects of dipole–dipole interaction by rotating a solid sample in a magnetic field. The rotation must be so rapid that its angular frequency ω_R is larger than the spread of Larmor frequencies in the sample. With proton resonance and a line width of only 1 gauss this method requires that $\omega_R/2\pi$ be 5000 Hz or greater. The averaging of the dipole–dipole interaction is maximized if the axis of rotation makes an angle $\theta = 54°44'$ with H_0. This is called the "magic angle" and it corresponds to $(1 - 3\cos^2\theta) = 0$. At this angle the only terms in Eq. (7.21) that are not averaged are those which are reduced to zero when $\theta_{ij} = 54°44'$. The "magic angle" corresponds to one-half the tetrahedral angle of 109°28′.

This kind of line narrowing can be very difficult because of the experimental problems associated with the large values of ω_R. It also removes the effects of the anisotropies in the spin–spin interaction and in the chemical shifts. One of the objectives in line-narrowing experiments in solids is to measure these anisotropies, which are lost in high-resolution solution spectra.

The most effective technique for narrowing the NMR spectra of solids is based upon pulse techniques. If the interacting nuclei are subjected to a selected series of short but intense H_1 fields, it is possible to obtain NMR signals that have been narrowed by the averaging of the dipole–dipole interactions. Since there is no crystal rotation, the anisotropies in chemical shifts and spin–spin interaction can also be directly measured. At the time of writing this book these experiments were just being fully developed, although the theory of these pulse experiments is well understood [see U. Haemberlen and J. S. Waugh, *Phys. Rev.*, **175,** 453 (1968)].

Figure 7.15 shows the results of one of these experiments. Since the sample used for the spectra shown in Fig. 7.15 was polycrystalline, the anisotropic parts of the spin–spin interaction and chemical shift are not resolved. The broad spectrum is the fully dipolar-broadened ^{19}F resonance in perfluorocyclohexane and it is about 1 gauss wide. The resolved AB quartet is obtained after a pulse-narrowing sequence has been used. This spin–spin pattern results from the axial and equatorial fluorines that make up each CF_2 group in the stable chair form of the

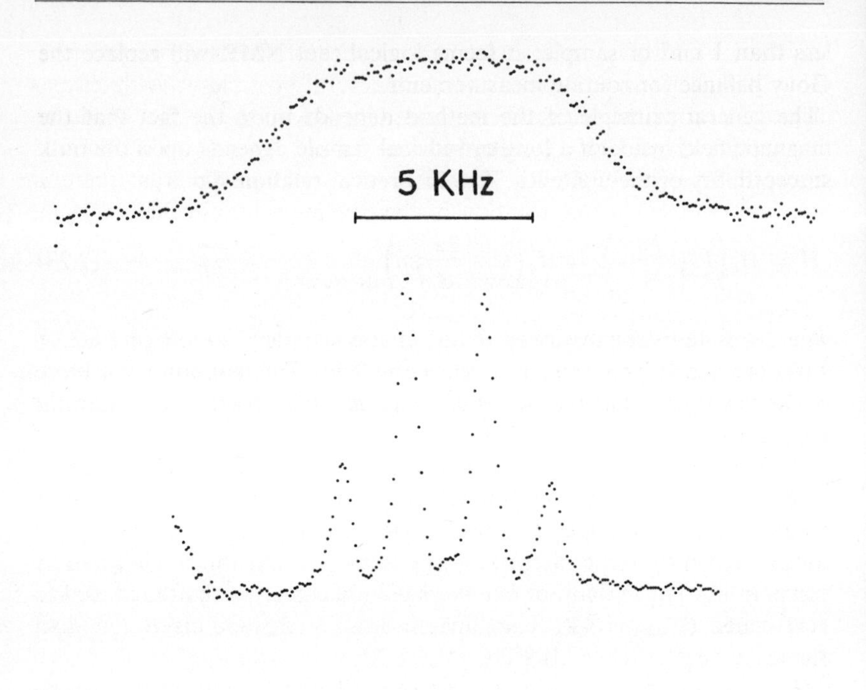

Figure 7.15. *Fourier transform of the* ^{19}F *transient nuclear resonance with pulse conditions to average dipole–dipole interaction. The broad resonance shows the normal dipole–dipole line width in a polycrystalline sample of* C_6F_{12}. *With pulse-narrowing techniques the axial and equatorial signals are resolved. Temperature* 200°*K and* $H_0 \approx 14$ *kG.* (*Courtesy J. Waugh.*)

cyclohexane ring. Although such pulse experiments require rather sophisticated electronic techniques, they show great promise for greatly increasing NMR resolution in solids.

7.5. Susceptibility Measurements by NMR

High-resolution NMR is very sensitive to the exact value of magnetic field present in the sample, and it is a convenient tool for the measurement of the bulk magnetic susceptibilities of liquids. Since many laboratories now possess commercial NMR instruments and they require less than 1 cm^3 of sample, it seems logical that NMR will replace the Gouy balance for routine measurements.

The general principle of the method depends upon the fact that the magnetic field inside of a long cylindrical sample depends upon the bulk susceptibility of the contents. The theoretical relationship is

$$H = H_0\left(1 + \frac{2\pi}{3}\kappa\right) = H_0\left(1 + \frac{2\pi}{3}\chi\rho\right) \tag{7.23}$$

where κ is the volume susceptibility of the sample. The factor $2\pi/3 = 2.094$ is a shape factor which would be zero for a sphere, but since most NMR spectra are taken in cylindrical sample tubes, Eq. (7.23) is generally applicable.

The method of D. F. Evans (1959) utilizes a regular 5-mm-diameter NMR tube filled with the solution of interest together with a small amount of dissolved reference material [2 per cent $(CH_3)_3COH$ for water solutions]. A similar concentration of reference material is dissolved in the pure solvent in a small capillary tube which is placed inside of the NMR tube. The two CH_3 resonance signals can be used in Eq. (7.16) to determine a δ value. This shift is related to the solvent density ρ_0 and mass susceptibility χ_0, and the solution density ρ_s and its mass susceptibility χ_s by Eq. (7.23) as

$$\chi_s = \frac{\chi_0\rho_0}{\rho_s} + \frac{3\delta}{2\pi\rho_s}\,. \tag{7.24}$$

If the solution were more paramagnetic than the solvent, its CH_3 signal would be less shielded than the solvent (away from TMS), and the δ term in Eq. (7.24) should have a positive value ($\chi_s > \chi_0$). The opposite should be true if $\chi_s < \chi_0$. If δ is in ppm, the susceptibilities will come out as $\chi \times 10^6$.

The Evans method with its internal standard depends upon the assumption that the reference material will not interact with the solute. Some paramagnetic ions may, for example, form a complex with $(CH_3)_3COH$ and give shifts that are not a measure of bulk properties. The method may also not work for two diamagnetic liquids since there are solvent effects to chemical shifts.

Both these problems are eliminated by the method of F. Frei and H. J. Bernstein (1962). In their method the reference material is always "external" to the solution and contained in both spherical and cylindrical tubes. A diagram of such an NMR cell is shown in Fig. 7.16. Since there should be no field shift for the spherical reference signal, one gets the same result as Eq. (7.24), if χ_s refers to the sample in the outer tube

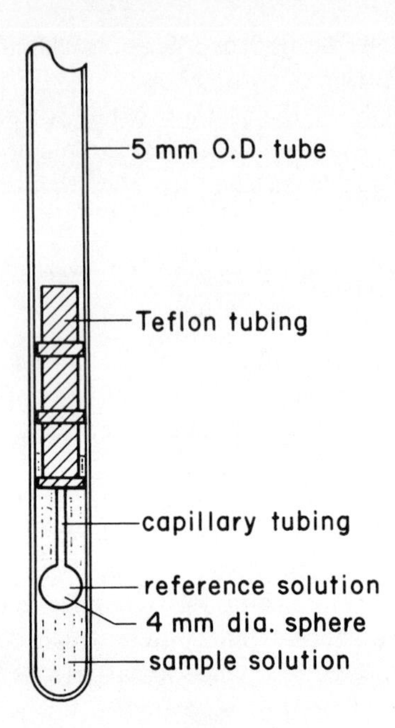

Figure 7.16. *NMR sample tube for susceptibility work. A modification of the design of L. N. Mulay and M. Haverbusch* [*Rev. Sci. Instr.*, **35,** 756 (1964).] *This design follows the method of Frei and Bernstein but utilizes a commercially available microcell. The tube must be positioned in the r.f. receiver coil of the spectrometer so that good NMR signals are observed from both the sphere and capillary containing the reference solution.*

and χ_0 to reference material in the inner tube. The shift δ, of course, only refers to the two NMR signals produced by the reference material in the inner tube. If the sample in the outer tube is more paramagnetic (on a volume basis) than the NMR reference in the inner tube, the spherical reference resonance will be at a lower field and act like a less shielded sample. If one follows the convention of assigning larger δ values to less shielded protons, we can write

$$\chi(\text{sample}) = \frac{\chi(\text{ref})\rho(\text{ref}) + 0.4775(\delta_{\text{sph}} - \delta_{\text{cyl}})}{\rho(\text{sample})}. \qquad (7.25)$$

Values for $\kappa(\text{ref}) = \chi(\text{ref})\rho(\text{ref})$ are given in Table 7.4 for some convenient reference liquids.

Table 7.4. *NMR Reference Liquids for Susceptibility Measurements*[a]

Liquid	$\chi(\text{ref})\,\rho(\text{ref}) \times 10^6$
Acetonitrile	−0.532
Tetramethylsilane	−0.543
Cyclopentane	−0.628
Water	−0.7217
Benzene	−0.799
Methyl iodide	−0.938
Bromoform	−0.948

[a] For 24°C. Data of K. Frei and H. J. Bernstein, *J. Chem. Phys.*, **37**, 1891 (1962), except for H_2O, benzene, and bromoform.

The factor $3/2\pi = 0.4775$ has been experimentally tested with liquids of known susceptibilities, and Frei and Bernstein obtained better agreement in their apparatus using 0.4860. For accurate work most authors recommend calibration of this factor, but for routine work with paramagnetic solutions this may not be necessary.

NMR methods have also been used on solid or powdered samples. In those NMR instruments with an external lock, large shifts can be easily measured with a coaxial tube arrangement and the external water sample lock as a reference [J. Q. Adams (1966)]. The tube inserts (see Fig. 7.16) are also convenient for chemical-shift work under conditions of changing susceptibilities. The data in Fig. 7.12 were taken with such an arrangement.

References

E. D. Becker, *High Resolution NMR, Theory, and Chemical Applications*, Academic Press, Inc., New York (1969). A short but complete introduction.

J. R. Dyer, *Applications of Absorption Spectroscopy of Organic Compounds*, Prentice-Hall, Inc., Englewood Cliffs, N.J. (1965). A nice little summary of high-resolution proton work on organic compounds.

J. W. Emsley, J. Feeney, and L. H. Sutcliff, *High Resolution Nuclear Magnetic Resonance Spectroscopy*, Vols. 1 and 2, Pergamon Press, Elmsford, N.Y. (1965). A rather complete summary of a vast amount of literature.

D. F. Evans, *J. Chem. Soc.*, **1959,** 2003; K. Frei and H. J. Bernstein, *J. Chem. Phys.*, **37,** 1891 (1962); J. Q. Adams, *Rev. Sci. Instr.*, **37,** 1099 (1966). The NMR methods for susceptibility measurements.

H. G. Hecht, *Magnetic Resonance Spectroscopy*, John Wiley & Sons, Inc., New York (1967). Additional details on high-resolution and broad-line NMR. At a reasonable level with a chapter on EPR.

J. A. Pople, W. G. Schneider, and H. J. Bernstein, *High-Resolution Nuclear Magnetic Resonance*, McGraw-Hill Book Company, New York (1959). The first important book written in the field.

Additional NMR references are given at the end of Chapter 6.

Problems

7.1. Calculate the difference for H_0 in gauss that will bring two protons into resonance that differ by one unit in δ. Assume constant frequencies of either 60 or 100 MHz.

7.2.* Calculate the difference in H_0 that will bring a proton spin–spin doublet into resonance which is split by 1 Hz.

7.3. Use the exact energy formula to calculate the spin–spin splitting in units of J for the case in Fig. 7.3 where $\nu_0(\sigma_a - \sigma_b) = 5J$.

7.4. The ^{19}F spectrum of SF_4 of $-100°C$ consists of two equal-intensity 1:2:1 triplets. The triplet splitting is 78 Hz and the two triplets have a separation corresponding to a chemical shift of 48 ppm. There is some additional structure on each member of the triplets at 56.4 MHz. What can you conclude about the molecular structure of SF_4 from these data? [See F. A. Cotton, J. W. George and J. S. Waugh, *J. Chem. Phys.*, **28,** 994 (1958) and J. Bacon, R. J. Gillespie and J. W. Quail, *Can. J. Chem.*, **41,** 1016 (1963).]

7.5. Construct a splitting diagram similar to Table 7.2 for one, two, three, and four spins with $I = 1$.

7.6. The proton spectrum of $GeAsH_5$ is a 1:2:1 triplet and a 1:3:3:1 quartet located at higher field. Predict a structure for this molecule. With this knowledge predict the proton spectrum of $GePH_5$. [See J. E. Drake and W. L. Jolly, *J. Chem. Phys.*, **38,** 1033 (1963).]

7.7. Iron pentacarbonyl has been considered to have either a pyramidal or bipyramidal structure. Since only one ^{13}C resonance has been observed in this compound, show how one can explain this fact. [See F. A. Cotton, A. Danti, J. S. Waugh and R. W. Fessenden, *J. Chem. Phys.*, **29,** 1427 (1958).]

7.8. The ^{19}F spectrum of PF_5 is sharp despite a possible broadening due to exchange between the axial and equatorial positions. Use Eq. (6.25) to calculate a lower limit to the lifetime of an ^{19}F in one position. Assume a maximum line width of 1 Hz and a chemical shift between the two positions of 60 ppm. Assume a frequency of 56.4 MHz. [See E. L. Muetterties, W. Mahler and R. Schmutzler, *Inorg. Chem.*, **2,** 613 (1963).]

7.9. The theory of nuclear quadrupole relaxation in liquids says that

$$\frac{1}{T_2} = \frac{3\pi^2(2I + 3)}{10I^2(2I - 1)} (eQq)^2\tau_c$$

where eQq is the quadrupole coupling constant in hertz. Use this equation to calculate the correlation time for tumbling for $TiCl_4$. This correlation time should be close to that predicted by Stoke's law,

$$\tau_c = 4\pi\eta a^3/3kT,$$

where η is the viscosity and a is the radius of a sphere. For $TiCl_4$ at 20°C assume that $\eta = 0.80 \times 10^3$ cgs and that $TiCl_4$ is a sphere of radius 4.0×10^{-8} cm.

7.10. Use Eq. (7.22) to calculate $(\Delta H^2)_{av}$ for a "powder" of SF_6 molecules that are far enough apart so that only intramolecular terms are important. Assume an octahedral geometry and an S—F bond distance of 1.58 Å. (See the text for an answer.)

7.11. Show that, in benzene substituted in the ortho and para positions by deuterium (benzene-2,4,6*d*), the intramolecular contributions to the proton $(\Delta H^2)_{av}$ are decreased by more than a factor of 2. This is then a good technique for separating intra- and intermolecular contributions in solid benzene. [See E. R. Andrew and R. G. Eades, *Proc. Roy. Soc.* (*London*) **A218,** 537 (1953) or Slichter.]

7.12.* With the Frei and Bernstein method, calculate the proton shift in parts per million if water is used as reference and benzene fills the outer tube. Repeat the calculation using methyl iodide as a reference.

Solutions:

7.2. For 60-MHz proton resonance, $H_0 \approx 14{,}000$ gauss, and with a 1-Hz splitting

$$\delta H \approx \frac{14 \times 10^3}{6 \times 10^7} \times 1 = 2.3 \times 10^{-4} \text{ gauss.}$$

7.12. With Eq. (7.25)

$$\delta_{\text{sph}} - \delta_{\text{cyl}} = 2.094(\chi_s\rho_s - \chi_0\rho_0).$$

For water as reference

$$\delta_{\text{sph}} - \delta_{\text{cyl}} = 2.094 \times (-0.077)$$
$$= -0.16 \text{ ppm.}$$

For methyl iodide as reference,

$$\delta_{\text{sph}} - \delta_{\text{cyl}} = 2.094 \times 0.139$$
$$= 0.29 \text{ ppm.}$$

In a 60-MHz instrument these would correspond -9.6 and 17.4 Hz, respectively.

8

Electron Paramagnetic Resonance

NUCLEAR MAGNETIC RESONANCE and electron paramagnetic resonance are very nearly complementary branches of spectroscopy. In diamagnetic systems containing nuclei with $I \geq \frac{1}{2}$, nuclear resonance can generally be observed. Electron paramagnetic resonance, on the other hand, requires a molecular magnetic moment and such moments form the basis of paramagnetism. Although it may be possible to observe NMR in systems that also possess molecular magnetic moments, as was illustrated in Chapter 7, the relaxation times that make it possible to observe NMR often make it difficult or impossible to observe EPR. In fact, we will find that EPR cannot be observed, under many experimental conditions, in some systems that are paramagnetic.

The total integrated intensity of a magnetic resonance spectrum depends upon the square of the interacting magnetic moment. Since the magnitude of the Bohr magneton is over 10^3 times that of the nuclear magneton, the intensity of an EPR spectrum can be over 10^6 times that of an NMR spectrum. It is possible to detect EPR, in favorable circumstances, in samples that contain as few as 10^{10}—or as little as 10^{-14} moles of—molecular magnetic moments. But, as we have indicated, in unfavorable circumstances it may not be possible to detect EPR in samples containing even 10^{20} moments. The relaxation times are the key to detectability, and short relaxation times can give very broad absorptions.

8.1. EPR Relaxation Times and Spin–Orbit Interaction

The width of a magnetic resonance signal can be related by the Bloch equations to the relaxation time T_2. Short T_2 values correspond to large widths, and when $1/T_2$ is close to the Larmor frequency the resonance is so broad that it cannot be detected. In electron resonance the times T_1 and T_2 can always be related to a fluctuating magnetic field H^*. In NMR, for example, the fluctuating field could usually be ascribed to the neighboring nuclear moments and it is only a few gauss in magnitude. In EPR one can usually dilute the sample with diamagnetic material and thereby reduce the effects of any neighboring electron moments to a very low level. However, in EPR the effects of spin–orbit interaction cannot be always so easily minimized.

In many atoms, spin–orbit interaction corresponds to energy splittings of 10^2 to 10^3 cm^{-1}. These splittings are equivalent to local magnetic fields of 10^6 to 10^7 gauss. In a molecule, the quenching of the orbital angular momentum reduces the effects of spin–orbit interaction but does not make it zero. In addition, the quenching of the orbital moment makes its contribution both anisotropic and sensitive to molecular vibrations. Under these circumstances relaxation times become very temperature dependent, with higher temperatures often giving shorter times.

The exact calculation of relaxation times in solids or liquids is a difficult problem. In solids the major source of electron spin relaxation is due to interactions between the electron spins and the vibrations of the solid lattice. Spin–orbit interaction is the mechanism for this relaxation. Its magnitude depends upon the vibrations of the lattice and it can be a source of a large fluctuating magnetic field. This was first proposed by W. Heitler and E. Teller (1936), who showed that the modulation of dipole–dipole interaction proposed by I. Waller (1932) was not important. At low temperatures in solids one usually has $T_2 \approx T_1$, and since T_1 is the easier quantity to calculate, most discussions of relaxation are primarily concerned with the origins of T_1. EPR measurements are commonly made using temperatures as low as 1°K and as high as 500°C. In this temperature range there can be a very considerable variation in the values of T_1 and T_2. If we start by considering temperatures of 100 to 300°K, we have kT values that are equal to the commonly observed spin–orbit splittings in transition-metal ions. Under these circumstances it is possible to make a simple generalization about the values of T_1. In systems with orbital degeneracy which can be split in first order by spin–orbit interaction, T_1 should be very short when kT is close to λ. These are also exactly those conditions which produce deviations from the Curie law. At temperatures as low as 1°K most orbital degeneracy is removed for the lowest state and the EPR spectra of most paramagnetic ions can then be detected.

Table 8.1. *Correlation of Electron Spin Relaxation Times with Orbital Degeneracy for Transition-Metal Ions in Octahedral Complexes*

Electron configuration	Ion	High spin term	Orbital degeneracy	Relative T_1 values
$(3d)^1$	Ti^{+3}	$^2T_{2g}$	3	short
$(3d)^2$	V^{+3}	$^3T_{1g}$	3	short
$(3d)^3$	Cr^{+3}	$^4A_{2g}$	1	long
$(3d)^4$	Cr^{+2}	^5E_g	2[a]	long
$(3d)^5$	Mn^{+2}	$^6A_{1g}$	1	long
$(3d)^6$	Fe^{+2}	$^5T_{2g}$	3	short
$(3d)^7$	Co^{+2}	$^4T_{1g}$	3	short
$(3d)^8$	Ni^{+2}	$^3A_{2g}$	1	long
$(3d)^9$	Cu^{+2}	^2E_g	2[a]	long

[a] Although these have a twofold orbital degeneracy, it is nonmagnetic in that spin–orbit interaction gives no first-order contribution to the energy, and in solids this permits long relaxation times.

Table 8.1 shows how T_1 varies for octahedral transition-metal complexes. One can see that for A states with no orbital degeneracy T_1 values are long. This is because the orbital angular momentum is well quenched for these states and spin–orbit interaction is not very large. The E states also have long T_1 values. This is because these E states have no first-order contributions from spin–orbit interaction and their g values are fairly close to the free electron. The T states, on the other hand, all have short T_1 values and their EPR spectra are not observed near room temperature.

The EPR spectrum of Fe^{+2}, for example, can be observed in octahedral MgO near 4°K but not at much higher temperatures. The Fe^{+3} spectrum, on the other hand, can usually be detected even at room temperature. Detailed theory for systems such as $Ti(H_2O)_6^{+3}$ shows that at a few degrees Kelvin T_1 varies by something like T^{-9}, and above a few degrees Kelvin one might expect to have only very short values for its T_1. From Table 3.2 one can see that $(3d)^5$ is reduced to a $^2T_{2g}$ state by large crystal fields. At the same time one finds that the EPR spectrum of Fe^{+3} in $Fe(CN)_6^{-3}$ is only detectable at low temperatures.

The symmetry of the crystal field is also very important. In V^{+4} or Ti^{+3} one finds very short T_1 values for high temperatures and crystal fields that are close to octahedral. In VO^{+2} there is a large axial crystal field. Its g values are close to the free-electron value and it follows the Curie law. As a result, the EPR spectra of VO^{+2} can easily be observed at room temperature. The second- and third-row transition-metal ions have still larger values for spin–orbit splittings and correspondingly shorter T_1 values. The lanthanide and actinide ions have orbital degeneracies in most cases and their EPR spectra cannot generally be observed except at very low temperatures.

In liquids, the relaxation times again depend primarily upon the nature of the spin–orbit interactions, although when $S > \frac{1}{2}$ the spin–spin interaction can also be very important. The tumbling in liquids tries to average out all anisotropic terms in the spin Hamiltonian, but such terms can still contribute to the line width as they do in NMR. In Table 8.1 one can see that Cr^{+3} complexes should be favorable since they can have no orbital degeneracy. However, for symmetries lower than octahedral the spin–spin splittings with $S = \frac{3}{2}$ are anisotropic, and some Cr^{+3} complexes have very broad EPR line widths in solution.

Most transition-metal complexes will have greater line widths in solution than they do in a solid single-crystal system. Under these circumstances one can expect a relatively small number of transition-metal ions to have sharp EPR spectra in solution. The largest source of sharp solution spectra are organic radicals. These radicals are only doublet states, they generally contain only atoms with small spin–orbit interactions, and they are covalent with no orbital degeneracy and g values that are close to the free-electron value. All these circumstances are favorable for yielding little spin relaxation for such radicals in solution.

8.2. Derivative Presentation in EPR

Most EPR instruments operate at fixed frequency. The magnetic field is slowly swept through the spectrum and at the same time it is modulated at frequencies from 10^2 to 10^5 Hz. The modulation is a small sinusoidal variation of H_0 up to a few gauss in amplitude. As a result of this modulation the signals presented by an EPR spectrometer are either the first derivative or even the second derivative of the absorption spectrum. At first glance this looks to be a disadvantage, for these derivative spectra are difficult to assimilate. There are, however, several distinct advantages

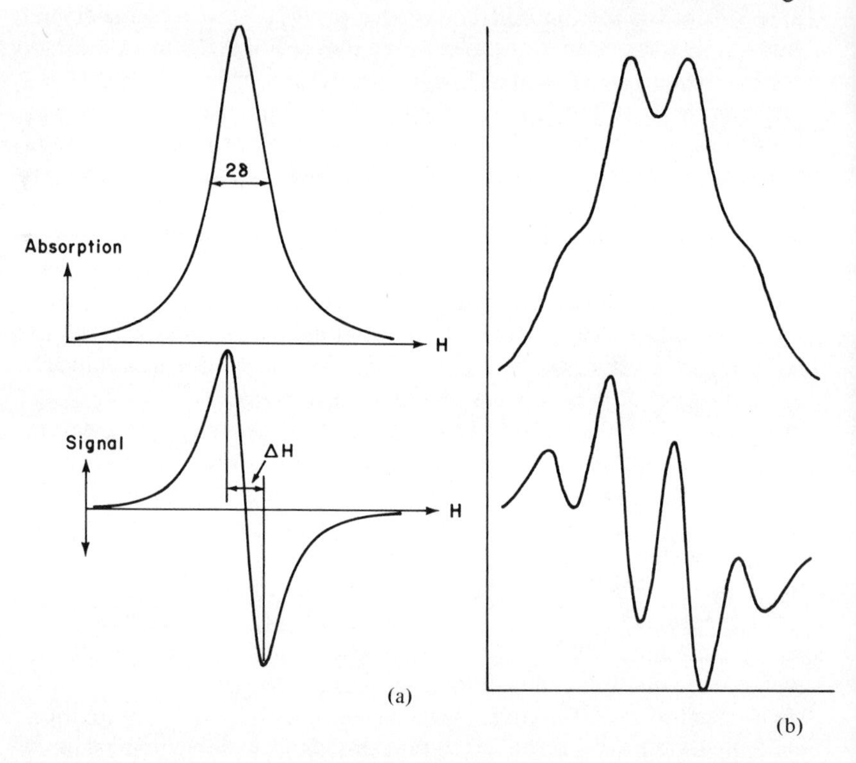

Figure 8.1. *Derivative presentation with Lorentz lines shapes. (a) A single absorption line and its first derivative. (b) Four equally spaced overlapping lines with* 1:3:3:1 *intensities and spacing equal to* 1.5δ.

to a derivative spectrum. Figure 8.1*a* shows a single absorption line and its first derivative.

Figure 8.1*b* shows a complex absorption spectrum of the type common to EPR. Also shown is its first derivative. One can see that the derivative spectrum seems to have greater resolution than the absorption spectrum. This is not really true, for all the derivative spectrum does is to accentuate the actual resolution in the original absorption spectrum. To the eye, however, the actual resolution is much more striking in derivative form, as is clearly shown in Fig. 8.1*b*.

Another advantage of the derivative spectrum is that it provides a well-defined line width. If ΔH is defined as the field difference between the minimum and maximum signals for the first derivative, this is certainly easy to measure from the derivative spectrum, as shown in Fig. 8.1*a*. Even if there are overlapping signals as shown in Fig. 8.1*b*, it is still possible to make good estimates of ΔH. A Lorentz absorption line has dropped to one-half its maximum value when $\omega - \omega_0 = 1/T_2$ or for a field sweep when $H - H_0 = 1/(\gamma T_2)$. The value of ΔH can be found by differentiation of the Lorentz line shape and one finds that $\Delta H = 2/(\sqrt{3}\gamma T_2)$. Table 8.2 lists some properties of the two common line-shape

Table 8.2. *Lorentzian and Gaussian Line-Shape Functions in Terms of Magnetic Fields*

	Lorentzian	Gaussian
Normalized line shape	$\dfrac{\delta/\pi}{\delta^2 + (H - H_0)^2}$	$\dfrac{e^{-(H-H_0)^2/2\sigma^2}}{\sigma\sqrt{2\pi}}$
$(H - H_0)$ for $\frac{1}{2}$ maximum	δ	$\sigma\sqrt{2 \ln 2} = 1.177\sigma$
$(H - H_0)$ for $\frac{1}{10}$ maximum	3δ	$\sigma\sqrt{2 \ln 10} = 2.146\sigma$
T_2	$\dfrac{1}{\gamma\delta}$	$\dfrac{\sqrt{\pi/2}}{\gamma\sigma}$ [a]
ΔH or $H_{max} - H_{min}$ in the derivative spectrum	$\dfrac{2\delta}{\sqrt{3}}$ or $\dfrac{2}{\sqrt{3}\gamma T_2}$	$\pi\sigma$ or $\dfrac{\sqrt{2\pi}}{\gamma T_2}$ [a]
$(\Delta H^2)_{av}$ or the second moment	∞	σ^2

[a] T_2 is only defined in the Bloch equations for a Lorentz line shape; T_2 can be defined as π/γ times the maximum in the normalized line-shape function. (See Andrew.)

functions. It can be seen that the Gaussian function drops off more rapidly than does the Lorentz function. This is made very striking when one sees that the second moment for a true Lorentz line is infinite. It is, of course, impossible to obtain an infinite second moment from an experimental line because of noise, so that experimental second-moment values are just large but never infinite.

8.3. Paschen–Back Hamiltonian for Spectra in Solution

For EPR spectra in solution the positions of the lines are determined by the isotropic terms in the spin Hamiltonian. The anisotropic terms would be exactly averaged to zero for short tumbling correlation times and they can contribute only to the line width. This can be expressed by

$$\mathsf{H}(\text{spin}) = \mathsf{H}(\text{isotropic}) + \mathsf{H}(t) \tag{8.1}$$

where $\mathsf{H}(t)$ is the time-dependent part and includes the anisotropic terms in the spin Hamiltonian. To be complete $\mathsf{H}(t)$ would also have to include the spin–phonon interactions, which are responsible for the line width in solids. The isotropic part of the spin Hamiltonian has been discussed previously and it contains an average g value and isotropic coupling terms between the odd electrons and the nuclei. The only isotropic part of electron spin–spin interaction is a spin-independent constant that we will neglect. For a number of nuclei one obtains

$$\begin{aligned}\mathsf{H}(\text{isotropic}) = {} & g\mu_B \mathbf{S} \cdot \mathbf{H}_0 \\ & - \hbar \sum_i \gamma_i \mathbf{I}_i \cdot \mathbf{H}_0 + \sum_i A_i \mathbf{S} \cdot \mathbf{I}_i ,\end{aligned} \tag{8.2}$$

where γ_i would include any chemical shifts for the nuclei.

This Hamiltonian is very similar to those previously considered for NMR and it was exactly solved in Chapter 2 for a doublet state and a single nucleus with $I = \frac{1}{2}$. Under ordinary EPR conditions the Paschen–Back effect is fairly complete and we can use the Z-axis terms in the dot products to form a first-order Hamiltonian. The result is

$$\mathsf{H}(\text{first order}) = g\mu_B H_0 \mathsf{S}_Z - \hbar \sum_i \gamma_i H_0 \mathsf{I}_{Z_i} + \sum_i A_i \mathsf{S}_Z \mathsf{I}_{Z_i} \tag{8.3}$$

where the magnetic field is taken as along the Z axis. The solution to Eq. (8.3) is

$$\epsilon(\text{first order}) = g\mu_B H_0 M_S - \hbar \sum_i \gamma_i H_0 M_{I_i} + \sum_i A_i M_S M_{I_i}. \tag{8.4}$$

The usual selection rules for EPR are

$$\Delta M_S = \pm 1, \qquad \Delta M_{I_i} = 0, \tag{8.5}$$

so that the second term in Eq. (8.4) cancels out and

$$\Delta\epsilon = h\nu = g\mu_B H_0 + \sum_i A_i M_{I_i}. \tag{8.6}$$

This result is rather similar to Eq. (7.14) for spin–spin splitting in NMR. In EPR the frequency is held constant and the field is swept. If we divide our nuclei into sets of equivalent nuclei A, B, etc., the fields for resonance are given by

$$H_0 = \frac{h\nu - A_a \sum_i^A M_{Ii} - A_b \sum_i^B M_{Ii} \cdots}{g\mu_B}. \tag{8.7a}$$

To form a splitting diagram we can use the splitting formed by each set R of equivalent nuclei,

$$(H_0^0 - H_0)_R = A'_R \sum_i^R M_{Ii} = A'_R M_{FR} \tag{8.7b}$$

where $H_0^0 = h\nu/g\mu_B$, and $A'_R = A_R/g\mu_B$ is the coupling constant in gauss. The number and relative intensity of the absorptions at $(H_0^0 - H_0)$ predicted by Eq. (8.7*b*) follow directly from methods used for spin–spin splitting in NMR. For $I = \frac{1}{2}$ the intensities will follow the binomial cofficients, and for $I > \frac{1}{2}$ a splitting diagram can be readily generated similar to Table 7.2. The simplest case is that of one nucleus of spin I, where Eq. (8.7) predicts $2I + 1$ lines, all with equal splitting and intensity.

Under extra high resolution or for very large A values the lines that are superimposed in the first-order spectrum can be split apart. In addition, the center of the spectrum is not exactly equal to $h\nu/g\mu_B$. When the second-order terms are included, the result for a single nuclear spin is

$$H_0 = \tfrac{1}{2}(H_0^0 - A'M_I) + \tfrac{1}{2}\sqrt{(H_0^0 - A'M_I)^2 - 2A'^2[I(I+1) - M_I^2]} \tag{8.8}$$

where $H_0^0 = h\nu/g\mu_B$ and $A' = A/g\mu_B$. For a single nuclear spin the result of the second-order term is to shift the center of the spectrum to lower field, and for even values of I

$$H_0(M_I = 0) = H_0^0 - \frac{A'^2 I(I+1)}{2H_0^0} + \cdots. \tag{8.9}$$

For rough measurements g is defined as $h\nu/\mu_B H_0$(center), but one can see from Eq. (8.9) that this is not very accurate for large A values.

8.4. Examples of EPR Spectra in Solution

The predictions of the first-order Hamiltonian are that these spectra should be symmetrical with uniformly spaced hyperfine splittings. Indeed, one can find many such examples in the literature. In this section we will discuss some examples showing both when simple spectra are observed and when they are not.

VO^{+2}, Mn^{+2}, AND Fe^{+3} IN H_2O

Figure 8.2 shows the spectrum of VO^{+2} in water near room temperature. Like most EPR spectra this one was taken at X band ($\nu \approx 9$ GHz) or $H_0^0 \approx 3000$ gauss. The eight lines arise from an isotropic hyperfine

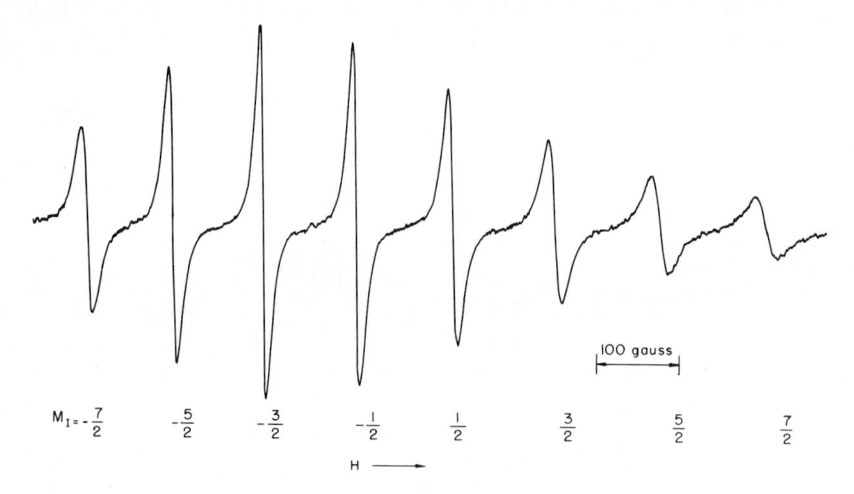

Figure 8.2. *EPR spectrum of* 0.01 *M* $VOSO_4$ *in water at room temperature. This spectrum was taken at 9 GHz and it is the first derivative of the absorption spectrum. EPR parameters:* $g = 1.962$, $A'(^{51}V) = 118$ *gauss.*

interaction with the 99.8 per cent abundant ^{51}V isotope which has $I = \frac{7}{2}$. The lines are nearly uniformly spaced as predicted by Eq. (8.7), but they do not seem to be of equal intensity in the derivative spectrum.

The difference between the eight lines shown in Fig. 8.2 is only due to their different line widths. Another advantage of the derivative spectrum is that it makes a difference in line width show up also as a difference in peak-to-peak amplitude. It is easier for the eye to detect a change in ΔH by the peak-to-peak amplitude than by the field shift between these two peaks. Except for the line-width variation with M_I the spectrum of VO^{+2} in water is exactly as expected from Eq. (8.7) or, more accurately, by Eq. (8.8).

A variation of line width with M_I can be explained by considering anisotropic terms in $\mathsf{H}(t)$ in Eq. (8.1), and this was first done for EPR by H. McConnell (1956). His theory is called the rotating microcrystallite theory, since it assumes an anisotropic Hamiltonian in solution which is related to that found in solids. In Chapter 6 we saw that one source of line width is a time-dependent Larmor frequency. The final result is summarized in Eq. (6.25). The anisotropic terms in the spin Hamiltonian become, in a liquid, a source of a time-dependent Larmor frequency. We can write

$$\hbar\,\delta\omega \approx (\delta g\,\mu_B H_0 + \delta A\,M_I) \tag{8.10a}$$

where δg is the anisotropy in g and δA is the anisotropy in the hyperfine coupling. When this is utilized in Eq. (6.25) we obtain

$$\frac{1}{T_2} \approx \tau_c(\delta g\,\mu_B H_0 + \delta A\,M_I)^2\hbar^{-2}. \tag{8.10b}$$

The correlation time in this equation would be that for the tumbling of the $VO(H_2O)_5^{+2}$ complex in solution. For water near room temperature it should be close to 10^{-11} seconds. If one considers all sources of relaxation, the general result could be expressed as a power series in M_I, so that

$$\frac{1}{T_2} = \alpha + \beta M_I + \gamma M_I^2 + \cdots \tag{8.10c}$$

where α, β, etc., should be dependent on both field and temperature. If Eq. (8.10*b*) represents the primary source of line width, one should have $\beta = 2\alpha\gamma$, and this is approximately true for the spectrum in Fig. 8.2, where γ is the most important term. One result of Eq. (8.10*b*) is that the ΔH values should all be smaller at higher temperatures where the correlation time for tumbling should be decreased. This is true for VO^{+2} in water solution, and this simple theory nicely accounts for the variation in line width with M_I.

The reason VO^{+2} has relatively sharp lines has been discussed earlier. The large axial component to its crystal field, along its z axis, removes the degeneracy associated with the V^{+4} ion in an octahedral field. The single $3d$ electron can be considered as in a well-stablized $d(xy)$ orbital which is split over 10,000 cm^{-1} below the $d(xz)$ and $d(yz)$. This is indicated by the closeness of g to the free-electron value.

The remaining question to the VO^{+2} spectrum concerns the observation of isotropic hyperfine splittings for ^{51}V when the odd electron is supposed to be in a $3d$ orbital. Fermi contact interaction is the primary source of isotropic hyperfine coupling, and it requires the odd electron to be partially in an s orbital so that its wave function at the nucleus is not exactly zero. A quantitative answer to this question is still not possible, but spin polarization of the $3s$ and $2s$ electrons is the qualitative answer.

In the spin polarization mechanism we take into account the fact that electrons of the same M_S value seem to have an attractive "exchange force" between them. This attractive "exchange force" can be also utilized to predict the Hund's rule stabilization of triplet states over singlet states by lowering the normal electron–electron repulsion.

The sign of the A value for VO^{+2} is very important. It is most probably negative, as the numbering of the M_I values for ^{51}V in Fig. 8.2 would indicate. A negative A value for ^{51}V means that the M_S value of the odd electron in the s orbital around the nucleus is opposite to that in the $3d$

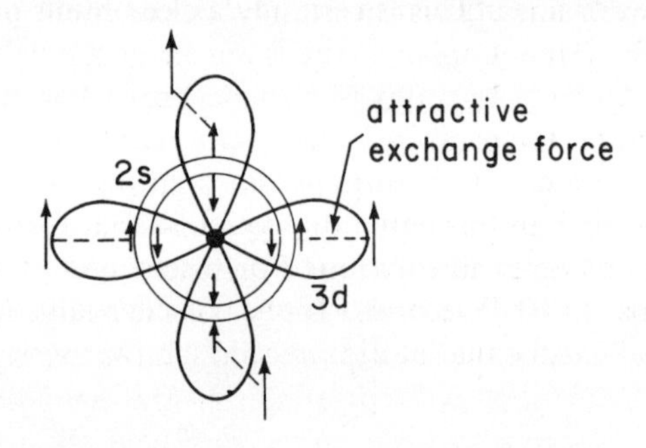

Figure 8.3. *"Exchange force" between the 3d and 2s electrons, with the same M_S value. The radius is smaller for the 2s electron with the opposite Ms value.*

orbital. Fig. 8.3 shows how this can be possible. An attractive "exchange force" between an outer $3d$ electron and an inner $2s$ electron makes the wave function of the $2s$ electron with the opposite sign of M_S have a larger value at the nucleus. For nuclei with positive nuclear magnetic moments this yields a negative A value, and while it is not possible from Fig. 8.2 to directly measure the sign of the A value for ^{51}V it is due to spin polarization and is most probably negative.

Both Mn^{+2} and Fe^{+3} should be favorable for EPR in solution since as $(3d)^5$ ions they correspond to $^6A_{1g}$ terms. The EPR spectrum of Mn^{+2} in water looks a good deal like Fig. 8.2, for the 100 per cent abundant ^{55}Mn has $I = \frac{5}{2}$ and so should give six equal lines. At room temperature one does observe six lines. They are spaced with an $A'(^{55}Mn)$ of about 95 gauss and with a ΔH for each line of about 25 gauss. At higher temperatures these lines narrow but at the same time they appear, like Fig. 8.2, to have different line widths.

At first glance this would seem to be another case for Eq. (8.10b), but in octahedral $Mn(H_2O)_6^{+2}$ both the g value and hyperfine splitting should be isotropic. The secret is now found in the failure of the Paschen–Back Hamiltonian for a $^6A_{1g}$ ion. With $S = \frac{5}{2}$ there are five possible EPR transitions: $M_S = -\frac{5}{2} \rightarrow -\frac{3}{2}$, $M_S = -\frac{3}{2} \rightarrow -\frac{1}{2}$, $M_S = -\frac{1}{2} \rightarrow \frac{1}{2}$, $M_S = \frac{1}{2} \rightarrow \frac{3}{2}$, and $M_S = \frac{3}{2} \rightarrow \frac{5}{2}$. In Eq. (8.7) these should all have the same field value for every M_I value. If one considers the second-order terms with such a large A value, these lines are not superimposed. The different line widths that can be observed for Mn^{+2} in solution are due to the different overlap of the five EPR transitions for every M_I value. This can be rapidly confirmed by lowering the microwave frequency to S band ($\nu \approx 3$ GHz), and the overlap is so bad that a six-line pattern cannot even be seen. On the other hand, the K band ($\nu \approx 24$ GHz) spectrum has six equal-width lines. This is certainly a clear result of the failure of the Paschen–Back Hamiltonian at the lower magnetic fields.

The spectrum of Fe^{+3} in water is a single broad line with ΔH about 1000 gauss at room temperature. There is no common Fe isotope with $I \neq 0$, so that there could be no hyperfine splitting. The line width for both Mn^{+2} and Fe^{+3} are primarily due to spin–spin zero-field splittings. A slightly distorted octahedron around the Mn^{+2} or Fe^{+3} ion can give a D value [see Eq. (4.7)] of nearly 0.1 cm^{-1}. This is a large interaction and although the mobility of the solution should tend to average this to zero, such dynamic distortions will contribute to the line width. It is well known that Fe^{+3} seems to have larger zero-field splittings than does Mn^{+2} in similar situations. Their solution-spectra line widths are clearly $\Delta H(Fe^{+3}) > \Delta H(Mn^{+2})$, in accordance with the expectations on the basis of D values arising from distorted octahedra.

If one adds a complexing agent to Mn^{+2} or Fe^{+3} in aqueous solution, their EPR spectra often seem to disappear. This is because these complexes are not octahedral, and since $S > \frac{1}{2}$, the complex will have a large D value in its spin Hamiltonian. Although this D value, for an ion in solution, is part of $\mathsf{H}(t)$, it can result in producing very broadened spectra. In solids, on the other hand, these zero-field splittings are not time dependent and such complexes can still give sharp EPR lines.

$Cr(CN)_5NO^{-3}$ IN H_2O

This ion is a good example of simple hyperfine splittings. In terms of crystal field theory it is best described as a complex containing CN^-, NO^+, and Cr^+. This makes it a $(3d)^5$ strong-field case with $S = \frac{1}{2}$. In an octahedral field it would have an unfavorable $^2T_{2g}$ ground state, but with its axial distortion the ground state has no orbital degeneracy and is best represented as $[d(xz), d(yz)]^4\, d(xy)$. The result of the removal of the orbital degeneracy is a rather sharp EPR spectrum, even in solution.

Figure 8.4 shows its EPR spectrum in water at room temperature. The strong 1:1:1 triplet arises from the $^{14}N(I = 1)$ of the NO^+. The moderately large A value for this ^{14}N must be due to a moderate delocalization

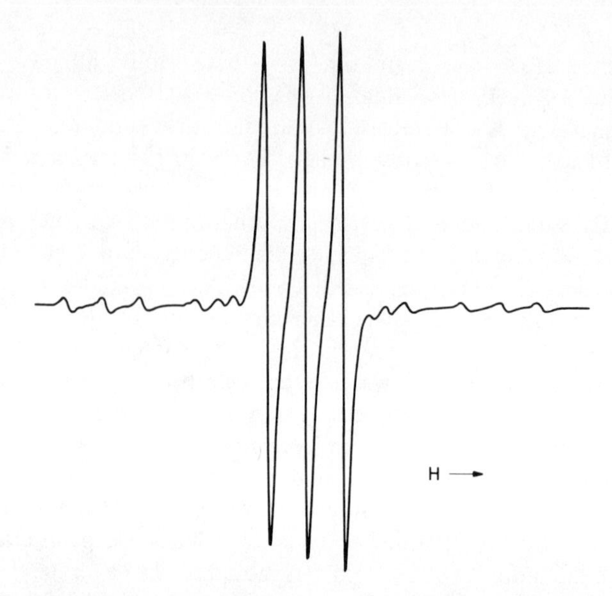

Figure 8.4. *EPR spectrum of* 0.01 *M* $K_3Cr(CN)_5NO$ *in water at room temperature. The lines are about* 2 *gauss wide and the EPR parameters are:* g = 1.9945, $A'(^{14}NO) = 5.26^*$, $A'(^{53}Cr) = 18.5$, $A'(^{13}CN$—*equatorial*) = 12.5*, $A'(^{13}CN$—*axial*) = 8.4* *all in gauss. The starred values are probably negative in sign.*

of the unpaired electron onto the 1*s* and 2*s* orbitals of the N in NO^+. This is clear evidence for the odd electron being in a molecular orbital and not only in a Cr^+ atomic orbital. The additional signals in Fig. 8.4 seem to form four sets of triplets. This obviously arises from the 9.5 per cent abundant ^{53}Cr ($I = \frac{3}{2}$) isotope. The other Cr isotopes all have $I = 0$.

Other than some small line-width variations the ^{14}N and ^{53}Cr structures all that is obvious from Fig. 8.4. Close examination of this figure shows that just outside the central lines there is an irregularity in the spacing of the ^{53}Cr signals. One should also be aware that the 1.1 per cent abundant ^{13}C ($I = \frac{1}{2}$) lines can have about the same signal strength as does ^{53}Cr because of the number of CN^- in the complex. Figure 8.5 shows the low-field portion of an EPR spectrum of a sample of $Cr(CN)_5NO^{-3}$ which has been enriched in $^{13}CN^-$. It is clear that the third line from the center in Fig. 8.4 has been greatly increased in Fig. 8.5 and that there is a new weak signal on the low-field end. The third line from the center is the outer line of a ^{13}C doublet of ^{14}N triplets from a $Cr(CN)_4{}^{13}CN^{14}NO^{-3}$ species, and the weak low-field signal is the outer line from a $^{53}Cr(CN)_4{}^{13}CN^{14}NO^{-3}$ species. All these splittings are consistent with the stated coupling constants and the first-order Hamiltonian. Extended

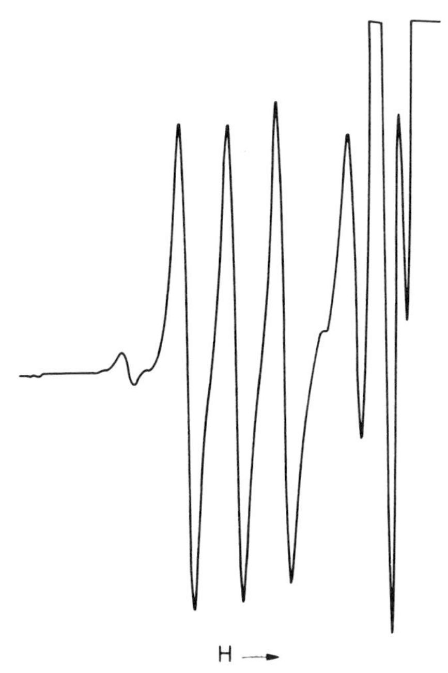

Figure 8.5. *Low field portion of the EPR spectrum of* $Cr(CN)_5NO^{-3}$ *enriched in* $^{13}CN^-$. *The first of the major triplets is shown as off scale to the far right.*

analysis of the ^{13}C splittings show that the clearly resolved lines belong to the more abundant equatorial $^{13}CN^-$ species, while a smaller splitting can be assigned to the $^{13}CN^-$ axial species. Splittings are not resolved for the ^{14}N of the CN^- groups, indicating much less delocalization of the unpaired electron past the carbon atom.

ORGANIC RADICALS

One of the more important uses of EPR has been the detection and assignment of the spectra of organic radicals. In solution these radicals give sharp lines and a rich hyperfine structure. Various techniques have been used for the generation of both neutral and charged radicals, including irradiation, electrolysis, chemical oxidation and reduction, and fast flow reactions. The small amount of radicals that are obtained by these techniques are compensated by the sensitivity of EPR. For most radicals in solution, good signals can be observed with a radical concentration of 10^{-4} to 10^{-5} molar. An important class of radicals are the ions that can be obtained by the one-electron oxidation or reduction of conjugated hydrocarbons. In these hydrocarbons the π system of electrons has relatively closely spaced energy levels and their first step one-electron oxidation or reduction products are π radicals.

Many properties of the π electrons in conjugated hydrocarbons can be well represented by the theory developed by E. Hückel (1931). In this theory the one-electron wave functions for the π system are represented by a linear combination of $2p_z$ orbitals on each of the n carbon atoms in the conjugated system. With this assumption

$$\psi_\pi(i) = \sum_n C_n(i)\psi_n(2p_z) \tag{8.11}$$

where $C_n(i)$ is a constant representing each carbon atom's participation in the i orbital of the π system. The constants $C_n(i)$ are determined by using a simplified effective Hamiltonian that we will not discuss. The Hückel Hamiltonian is so simple, however, that all the $C_n(i)$ can be readily determined for any conjugated hydrocarbon without the use of either a physical constant or an adjustable parameter. In a radical anion the odd electron is assumed to be in the lowest unoccupied Hückel orbital and in a radical cation it is in the highest occupied Hückel orbital, of the conjugated hydrocarbon. This is illustrated in Fig. 8.6.

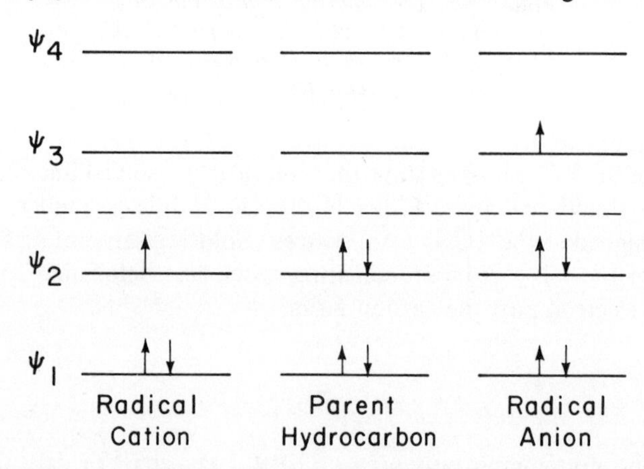

Figure 8.6. *Hückel energy levels for* 1,3-*butadiene. The dotted line is the zero on the energy scale and it corresponds to the Coulomb integral* α *in Hückel theory. The levels are spaced at* $\pm 0.618\beta$ *and* $\pm 1.618\beta$ *where* β *is the resonance or bond integral.*

The simplest conjugated hydrocarbon is 1,3-butadiene. It has four electrons in its π system and these would occupy the first two Hückel orbitals. The Hückel coefficients for this molecule are given in Table 8.3, and one can see that they form a very simple pattern. The radical

Table 8.3. *Hückel Coefficients for* 1,3-*Butadiene*

Orbital[a]	Coefficient[b]			
	C_1	C_2	C_3	C_4
1	0.3717	0.6015	0.6015	0.3717
2	0.6015	0.3717	−0.3717	−0.6015
3	0.6015	−0.3717	−0.3717	0.6015
4	0.3717	−0.6015	0.6015	−0.3717

[a] In order of increasing energy.
[b] For carbon atoms numbered in bonding order.

anion of 1,3-butadiene has been prepared and its odd electron can be approximated as occupying the third Hückel orbital. Before we relate these Hückel coefficients to the properties of the radical anion, we should first assign its EPR spectrum shown in Fig. 8.7.

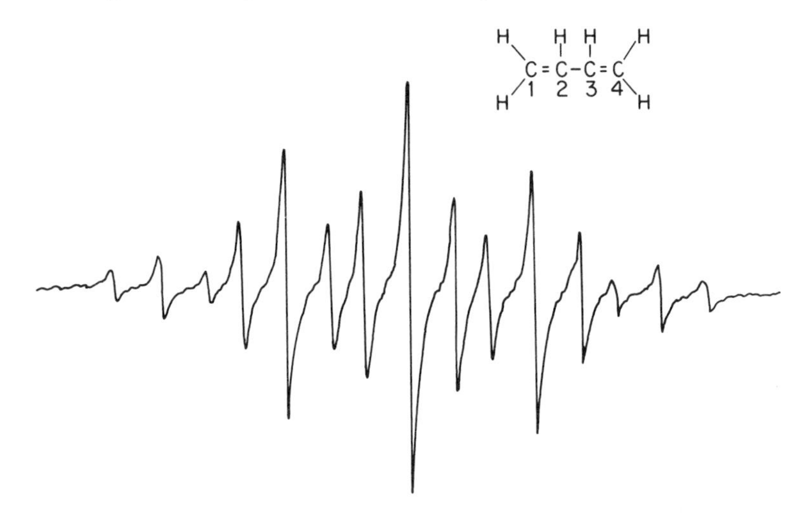

Figure 8.7. *EPR spectrum of the radical anion of* 1,3-*butadiene. The coupling constants are* $A_1'(H) = 7.617 \pm 0.005$ *and* $A_2'(H) = 2.791 \pm 0.007$ *gauss where both values are presumed to be negative in value.* [*From D. H. Levy and R. J. Myers, J. Chem. Phys.*, **41**, 1062 (1964).]

Butadiene has four hydrogens bonded to the end carbons and two bonded to the middle ones. In an *s*-cis or *s*-trans conformation the four hydrogens are not exactly equivalent, but they are all equivalent with respect to the π system. The first-order Hamiltonian result in Eq. (8.7) predicts a hyperfine splitting with the intensities of 1:4:6:4:1 for the four end hydrogens and 1:2:1 for the two middle ones. One can see from Fig. 8.7 that the quintet splitting is larger than the triplet one, so that it appears as a quintet of triplets. All the lines have the same width and they all have the proper intensities.

The coupling constants given in Fig. 8.7 are primarily due to Fermi contact interaction and must be caused by the odd electron spending some time on the 1*s* wave function of each hydrogen. It is no small theoretical problem to see how an electron that is primarily in a $2p_z$ orbital on a carbon can utilize the 1*s* orbital on each hydrogen. This is, however, very similar to the related problem in transition-metal ions for getting from a 3*d* orbital to 2*s* and 3*s* orbitals on the central ions. Both effects are a result of electron–electron repulsion and they can be explained by the same "exchange force." Figure 8.8 shows how this

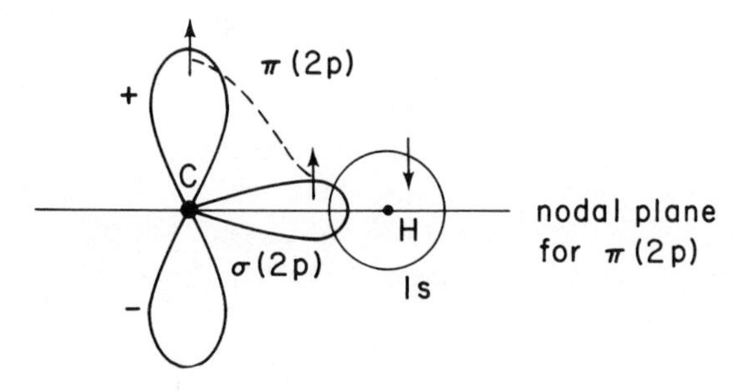

Figure 8.8. *Spin polarization of the* C—H *bond by a* $\pi(2p_z)$ *electron. The attractive "exchange force" between electrons of the same* M_S *value forces the* 1*s orbital on the hydrogen to have a greater probability of an electron with the opposite* M_S *value.*

"exchange force" pulls the electron with the same M_S value, in the C—H bond, toward the carbon atom. The result is an excess of electron spin with the opposite M_S value left on the 1*s* hydrogen orbital, so that the $A(H)$ values should be negative in sign.

H. McConnell (1956) proposed a linear relation in conjugated hydrocarbons for this spin polarization. The $A(H)$ value is proportional to the probability that the electron is on the 1*s* hydrogen orbital, and the square of the Hückel coefficient is the probability that the electron is in a given $2p_z$ orbital. McConnell's relation is that

$$A'_n(H) = QC_n^2(i) \tag{8.12}$$

where $C_n(i)$ is the Hückel coefficient for the carbon atom forming the C—H bond and Q is the necessary constant in gauss. By both theory and experiment it is found that $Q \approx -25$ gauss. The result of this relation for butadiene is that we should have

$$\frac{A_1'(H)}{A_2'(H)} = \frac{C_1^2(i)}{C_2^2(i)}, \tag{8.13}$$

and for the radical anion these should be the coefficients in Table 8.3 for the third Hückel orbital. The coefficients give $C_1^2/C_2^2 = (0.6015)^2/(0.3717)^2 = 2.618$, while the experimental coupling constants give $A_1'(H)/A_2'(H) = 7.617/2.791 = 2.729$. The agreement between these two numbers is excellent. In addition, we have

$$\sum_n A_n'(H) = Q \sum_n C_n^2(i) = Q\,, \tag{8.14}$$

so that our experimental Q value is -20.82 gauss where we assume our coupling constants to be all negative in value.

The result of this simple theory for 1,3-butadiene is to accurately predict the experimental radical-anion coupling constants with a single parameter Q. For most of the conjugated hydrocarbons one finds that similarly good agreement can be obtained between theory and experiment. Such agreement has been one of the high points of agreement in physical chemistry between simple theory and accurate spectroscopic results.

The radical cation of butadiene has never been examined in EPR, but a number of aromatic hydrocarbons have fairly stable radical cations. For the larger aromatics one simply has to dissolve the hydrocarbon in concentrated H_2SO_4. Figure 8.9*a* shows the EPR spectrum of the radical cation of tetracene. This hydrocarbon has three sets of hydrogens with four hydrogens in each set. Its EPR spectrum should have $5 \times 5 \times 5 = 125$ lines, or a series of interwoven quintets. Figure 8.9*a* clearly shows about one-half the possible lines. Since the strongest lines have $6 \times 6 \times 6 = 216$ times the intensity of the weaker ones, it is not too surprising that all lines cannot be seen in the same spectrum. Figure 8.9*b* shows a theoretical plot of the spectrum with the line width and coupling constants given in the figure. Another reason that fewer than 125 lines can be detected is that about 40 of the lines cannot be resolved since one coupling constant is close to being an exact submultiple of another.

An interesting feature of the spectrum given in Fig. 8.9*a* is that rather low microwave power had to be used. This is because the lines are rather narrow and T_2 is large. Since H_2SO_4 is rather viscous, T_1 is also large. The result is a relatively large T_1T_2 product and possible power saturation as given by Eq. (6.18). The behavior of T_1 and T_2 with rotational correlation time follows the form given by Eq. (6.14). Most solution EPR spectra are run under conditions where $\tau_c \approx 1/\omega_0$, because

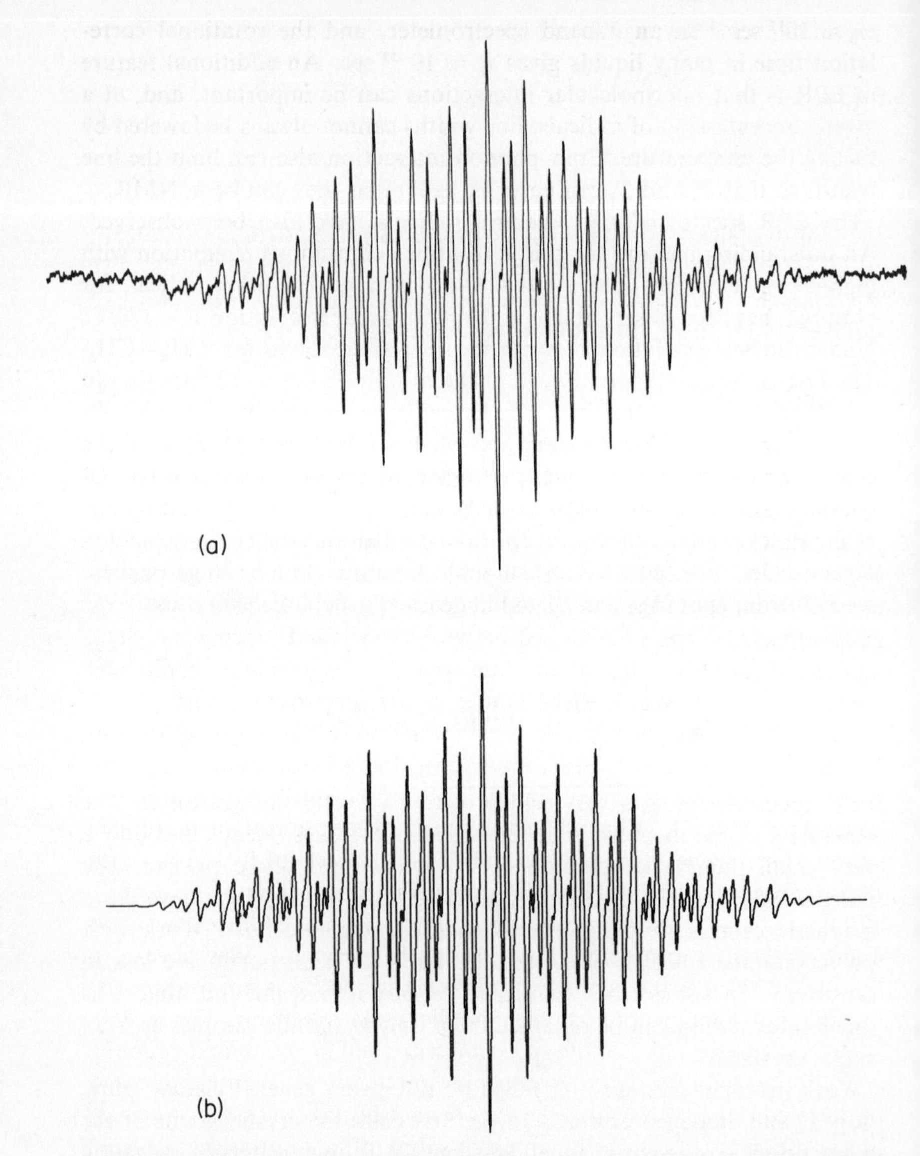

Figure 8.9. *EPR spectrum of tetracene radical cation. (a) The experimental spectrum obtained from* 10^{-4} *M solution of tetracene in* H_2SO_4 *at* 70°*C. (b) A theoretical spectrum with* $\Delta H = 0.15$ *gauss and coupling constants for three sets of four hydrogens each of* 1.03, 1.69, *and* 5.06 *gauss.*

$\omega_0 \approx 10^{11}$ sec^{-1} in an *X*-band spectrometer, and the rotational correlation time in many liquids gives $\tau_c \approx 10^{-11}$ sec. An additional feature in EPR is that intermolecular interactions can be important, and, at a given concentration of radicals, line widths cannot always be lowered by raising the temperature. Spin–phonon interaction also can limit the line width, so that T_1 and T_2 can never be as long as they can be in NMR.

The EPR spectra of many neutral radicals have also been observed. An outstanding method for this is to utilize continuous irradiation with high-energy electrons into a liquid hydrocarbon. The ethyl radical, for example, has been observed in liquid ethane under irradiation at -170°C. Under the best resolution 24 hyperfine lines are resolved for $\cdot CH_2—CH_3$. The first-order result from Eq. (8.7) is that only $3 \times 4 = 12$ lines should be observed. The extra lines can be explained by adding the second-order terms to the Hamiltonian, and what one finds is a doubling of the center of each triplet and a quadrupling of the center of each quartet. Of course, a single transition can never be split by removing the restrictions of the Paschen–Back effect, but multiple transitions can usually be split in higher order. For the ethyl radical with A' values close to 20 gauss these second-order splittings are 0.1 to 0.3 gauss at a field of 3000 gauss.

8.5. EPR Spectra in Solids

EPR spectroscopy is a very powerful tool in solid-state research. The sensitivity of the method, in solids with narrow lines, means that only a very small number of paramagnetic centers need to be present. The narrow lines are best obtained in single-crystal solids which have paramagnetic centers properly diluted in a diamagnetic host. Work with polycrystalline solids is also possible, but with a proportionate loss in sensitivity. In some cases, sensitivity is not important and almost as much information can be obtained from polycrystalline samples as from single crystals.

Work in single-crystal solids falls into three very general classes: pure, doped, and damaged crystals. In the first class the crystal is one of the many possible pure paramagnetic systems. Although waters of hydration and bulky anions may prevent serious spin pairing in such crystals, their EPR spectra can have rather broad lines. This amounts to a loss of resolution and so most investigators prefer to work in doped crystals.

In a doped crystal a diamagnetic host is grown, or found, with the proper level of paramagnetic impurity. To ensure good isolation the ratio of diamagnetic host to paramagnetic center is usually higher than 100:1. Since the site symmetry for the paramagnetic center is largely determined by the site symmetries of the diamagnetic host, the hosts are chosen with particular care. The M^{+2} sites in MgO, for example, are octahedral with sixfold coordination while the M^{+2} sites in CaF_2 (fluorite) are cubic with eightfold coordination. In some cases the substitution of an ion or molecule in a lattice can lead to a distortion. For this reason it is possible for the paramagnetic center to have a lower symmetry than the original site. EPR spectra can also be particularly sensitive to such small distortions from ideal symmetry.

A crystal can be damaged by a variety of processes. The simplest are either heat or light, but both x rays and neutrons can be very effective at producing paramagnetic centers in crystals. A great deal of work has been done with EPR in studying the radicals that can be produced in organic crystals by radiation damage. The resultant radicals can have fairly sharp EPR lines, even at room temperature, and a detailed hyperfine structure can usually be resolved. In some cases a number of radicals are produced, and considerable time can be spent trying to determine their formulas and structures from the *g* values and hyperfine splittings.

Polycrystalline studies can be done using the same preparative techniques as those exploited for single crystals. In addition, one may form glasses by the rapid freezing of gases or liquids. A number of free radicals can be prepared in polycrystalline form by freezing the radicals rapidly from the gas phase along with an inert matrix material like Xe or Ar. To prevent the slow diffusional recombination of the radicals in the matrix it is necessary to drop the temperature very far below the freezing point of the matrix. Matrix work using the inert gases is done at the boiling point of hydrogen (20°K) or of helium (4°K). If some liquid solutions are rapidly frozen in liquid nitrogen, the contents of the solution will be included in essentially a glass structure. This results in the paramagnetic molecules being oriented in all directions in space. The EPR spectrum of such a random system can be relatively simple, and all the parameters in the spin Hamiltonian can often by completely determined from such spectra.

The major characteristic of EPR spectra in solids are the effects of the anisotropies in the *g* values and in the hyperfine and the electron spin–spin interactions. The anisotropies in the *g* values, and in the hyperfine interaction, depend upon the nature of the crystal or ligand fields. Such anisotropies can be calculated with reasonable accuracy from simple theories. The anisotropy in the spin–spin interaction, which exists for triplet and higher multiplicities, is more difficult to calculate since it depends upon the relative positions of two electrons. All these effects are well represented, however, in the spin Hamiltonian, and an EPR spectrum can be completely assigned without any explicit use made of electron wave functions.

$Cu^{+2}[(CH_3COCHCOCH_3)^{-1}]_2$

Copper acetylacetonate forms a rather simple chelate structure. As shown in Fig. 8.10 the four carbonyl oxygens form a nearly rectangular crystal field in an essentially square planar complex. The closeness of the crystal field to full fourfold symmetry means that the EPR spectrum

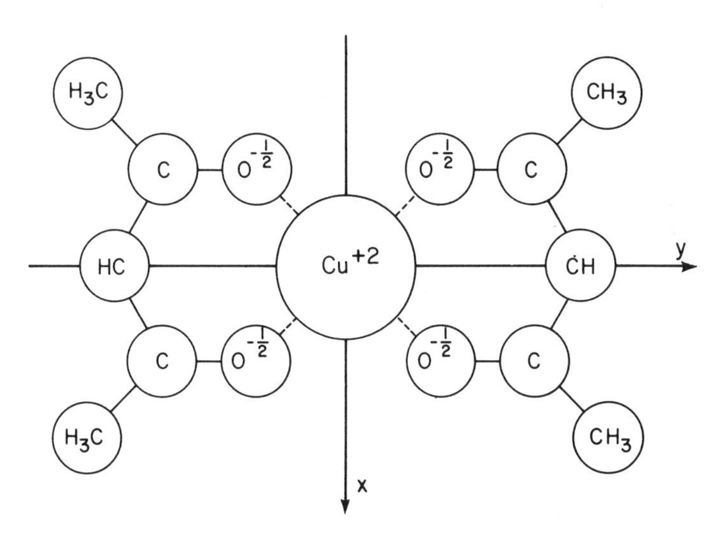

Figure 8.10. *Geometry of copper acetylacetonate. The oxygens form a square planar complex around the* Cu^{+2}. *Note the 45° rotation of the x and y axes from those used for octahedral symmetry.*

might be well represented by an axial spin Hamiltonian. With only nuclear hyperfine interaction from the copper nucleus the necessary spin Hamiltonian is

$$\mathsf{H} = \mu_B g_{\|} S_z H_z + \mu_B g_{\perp}(S_x H_x + S_y H_y) + A_{\|} S_z I_z + A_{\perp}(S_x I_x + S_y I_y), \tag{8.15}$$

where we have added hyperfine terms to Eq. (4.2). Two additional terms can also be added to Eq. (8.15): one for nuclear quadrupole interaction and one for the nuclear Zeeman effect. In the final analysis neither of these is very important in the assignment of the spectrum. With the selection rules $\Delta M_S = \pm 1$ and $\Delta M_I = 0$ the transitions allowed by Eq. (8.15) can be calculated to first order by

$$H_0(\theta) = \frac{h\nu - A(\theta)M_I}{\mu_B g(\theta)}$$

where

$$g(\theta) = (g_{\parallel}^2 \cos^2 \theta + g_{\perp}^2 \sin^2 \theta)^{1/2}$$

and

$$A(\theta) = \frac{(A_{\parallel}^2 g_{\parallel}^2 \cos^2 \theta + A_{\perp}^2 g_{\perp}^2 \sin^2 \theta)^{1/2}}{g(\theta)}. \tag{8.16}$$

The angle θ is that between the applied field H_0 and the z axis of the complex. One should not confuse these $A(\theta)$ and $g(\theta)$ values with those used for the Hamiltonian in solutions for

$$\begin{aligned} A(\text{solution}) &\equiv \langle A \rangle_{\text{av}} = \frac{A_{\parallel} + 2A_{\perp}}{3} \\ g(\text{solution}) &\equiv \langle g \rangle_{\text{av}} = \frac{g_{\parallel} + 2g_{\perp}}{3}, \end{aligned} \tag{8.17}$$

and these solution values are different from the ones used in Eq. (8.16). While Cu^{+2} has two common isotopes, ^{63}Cu and ^{65}Cu, they both have $I = \frac{3}{2}$. They also have sufficiently close magnetic moments so that they can ordinarily both be treated with a single set of hyperfine coupling constants.

A. H. Maki and B. R. McGarvey (1958) doped the diamagnetic Pd^{+2} acetylacetonate with 0.5 per cent of the Cu^{+2} complex. The two structures form isomorphorus crystals that are monoclinic but have only two inequivalent molecules per unit cell. In a single crystal spectrum four hyperfine lines will be produced for each molecule in the unit cell, and an arbitrary orientation of H_0 along the crystal axes will give eight possible lines. From such spectra Maki and McGarvey found the g and A values listed in Table 8.4.

Since Cu^{+2} has a $(3d)^9$ configuration, it corresponds to one hole in a filled $(3d)^{10}$. A square planar complex is equivalent to a large tetragonal distortion along the z axis of an octahedral complex. From a comparison of Figs. 4.2 and 3.6 one can see that the hole, or the odd electron, should be in a $d(x^2 - y^2)$ orbital. However, the coordinate system shown in Fig. 8.10 is rotated by 45° in the xy plane from that used for octahedral complexes. With such a rotation the odd electron should have its lowest energy in what one now calls the $d(xy)$ orbital. This is because it is this orbital which now points directly at the oxygen in Fig. 8.10. If one considers the molecular orbitals in the ligand field method, this coordinate system allows the $d(xz)$ and $d(yz)$ orbitals to form out-of-plane π

Table 8.4. *EPR Parameters Found for Copper(II) Acetylacetonate*

Parameter	Method	
	single crystal[a]	glass[b]
$g_{\parallel}$	2.266	2.264
$g_{\perp}$	2.054	2.036
$A_{\parallel}/hc$[c]	-160×10^{-4} cm^{-1}	$\pm 146 \times 10^{-4}$ cm^{-1}
$A_{\perp}/hc$	-20×10^{-4} cm^{-1}	$\pm 29 \times 10^{-4}$ cm^{-1}

[a] In crystals of Pd(II) acetylacetonate, see A. H. Maki and B. R. McGarvey, *J. Chem. Phys.*, **29**, 31 (1958).

[b] In glass of 60 per cent toluene and 40 per cent chloroform; see H. R. Gersmann and J. D. Swalen, *J. Chem. Phys.*, **36**, 3221 (1962).

[c] A(gauss) $= A/g\mu_B$ and $A(10^{-4}\text{ cm}^{-1}) = A/hc \times 10^4$. Since $\mu_B/hc = 4.67 \times 10^{-5}$, one finds that $A(\text{gauss}) \approx A(10^{-4}\text{ cm}^{-1})$ when $g \approx 2$. The exact relation is given in Section A.2.

bonds with a pair of oxygen atoms in each acetylacetonate. At the same time the $d(x^2 - y^2)$ forms an in-plane σ bond and the $d(xy)$ forms an in-plane π bond with the orbitals on these oxygens. For these reasons this coordinate system seems to be better than the usual one at 45°, and it clearly places the odd electron in a $d(xy)$ orbital. For this kind of complex the ligand field approach is probably better than the simple crystal field model, but we will explain the g and A values using the simpler crystal field equations.

In the crystal field model we can write

$$g_z = g_{\parallel} = 2.0023 - 2\lambda\Lambda_z$$
$$g_x = g_y = g_{\perp} = 2.0023 - 2\lambda\Lambda_x \tag{8.18a}$$

where

$$\Lambda_i = \sum_{n \neq 0} \frac{|\langle\psi_0^0|\, \mathsf{L}_i \,|\psi_n^0\rangle|^2}{\epsilon_n^0 - \epsilon_0^0}. \tag{8.18b}$$

For Cu^{+2} in a large tetragonal field we can limit the summation in Eq. (8.18*b*) to the single manifold of crystal field levels, and one can show that

$$\Lambda_z = \frac{4}{\epsilon^0(x^2 - y^2) - \epsilon^0(xy)} = \frac{4}{\Delta_1}$$

and

$$\Lambda_x = \frac{1}{\epsilon^0(xz, yz) - \epsilon^0(xy)} = \frac{1}{\Delta_2}. \tag{8.19a}$$

If we use these values in Eq. (8.18a), the result is

$$g_{\parallel} = 2.0023 - \frac{8\lambda}{\Delta_1}$$
$$g_{\perp} = 2.0023 - \frac{2\lambda}{\Delta_2}. \tag{8.19b}$$

Spectroscopic measurements on Cu^{+2} acetylacetonate show that $\Delta_1 \approx$ 15,000 cm^{-1} and $\Delta_2 \approx 25{,}000$ cm^{-1}. With these values in Eq. (8.19b) the single-crystal g values in Table 8.4 give $\lambda = -500$ and -650 cm^{-1}.

The two values calculated for the spin–orbit constant λ for Cu^{+2} are both smaller in magnitude than the -852 cm^{-1} listed in Table 3.1. This is a common type of failure of the crystal field model, and spin–orbit constants reduced to 60 to 80 per cent of the free-ion value are commonly used. Ligand field expressions based upon molecular orbitals can be substituted for Eq. (8.19b). In these equations one finds that the reduced spin–orbit interaction is a direct result of the fact that the unpaired electron spends a fraction of its time on the oxygen orbitals. Reduce spin–orbit interaction is a direct result of the participation of the oxygen atom in the orbitals for the unpaired electron.

The hyperfine interaction has two sources. The Fermi contact interaction is isotropic and it should contribute equally to $A_{\parallel}$ and $A_{\perp}$. The second source is the dipole–dipole interaction between the electron and the nucleus, and it accounts for the difference between $A_{\parallel}$ and $A_{\perp}$. This dipole–dipole interaction can be calculated by using the equations in Fig. 5.5, if one considers the magnetic energy of the electron in the field produced by the nucleus. From Fig. 5.5 one can see that this energy should drop off as r^{-3} and it should also depend upon the angular distribution of the electron. The two kinds of terms for an electron in a pure $d(xy)$ orbital yield the result,

$$A_{\parallel} = P(-\tfrac{4}{7} - \kappa) \tag{8.20a}$$

$$A_{\perp} = P(\tfrac{2}{7} - \kappa) \tag{8.20b}$$

where

$$P = 2.0023 g_I \mu_N \mu_B \left\langle \frac{1}{r^3} \right\rangle. \tag{8.20c}$$

The first terms in both Eqs. (8.20a) and (8.20b) are the dipole–dipole interactions where $\frac{2}{7}$ is the result of the average over the angular variation for the $d(xy)$ function. The additional factor of -2 between the first terms in $A_{\parallel}$ and $A_{\perp}$ can be explained by the fact that $A_{\parallel}$ refers to the energy when H_0 is along z and $A_{\perp}$ to when H_0 is in the xy plane. In each case the electron is in the $d(xy)$ orbital. This same factor of -2 is found in Fig. 5.5 between the values for H(dipole) for $\theta = 0$ and $\theta = 90°$. The Fermi contact term is $-P\kappa$, where the negative sign is a natural result of the spin polarization mechanism. With these equations one has A(solution) only dependent upon the Fermi contact interaction and

$$A(\text{solution}) = \frac{A_{\parallel} + 2A_{\perp}}{3} = -P\kappa. \tag{8.21}$$

These hyperfine equations are incomplete, for they neglect the orbital contributions to the magnetic moment of the electron. It is these orbital contributions that make the g values differ from the free-electron value, and for Cu^{+2} they are obviously important for $g_{\parallel}$. The major orbital contribution is to $A_{\parallel}$, and one obtains

$$A_{\parallel} = P[-\tfrac{4}{7} - \kappa + (g_{\parallel} - 2.0023) + \cdots]. \tag{8.22}$$

One interesting aspect of the use of Eq. (8.22) is that A(solution) is no longer due only to Fermi contact interaction. This is reasonable, for the dipole–dipole terms should only average out to zero in solution if the magnitudes of the dipole moments are independent of orientation. The orbital moment is itself anisotropic, and so it provides a contribution to A(solution). This contribution is often called "pseudocontact interaction," and in most cases it is much smaller than the observed Fermi contact interaction.

For $^{63}Cu^{+2}$ good theoretical calculations of $\langle 1/r^3 \rangle$ for a $3d$ electron show that $P/hc = 400 \times 10^{-4}\ \text{cm}^{-1}$. If we use Eqs. (8.20*b*) and (8.22) for copper acetylacetonate, they predict that $A_{\parallel} - A_{\perp} = -P(\tfrac{6}{7} - 0.264) = -240 \times 10^{-4}\ \text{cm}^{-1}$. This value is a little too large in magnitude but still close to the values in Table 8.4. In a molecular-orbital approach the electron spends part of its time on the orbitals of the oxygen. During this time the dipole–dipole interaction between the electron and the ^{63}Cu should be essentially zero. In this way one can see that delocalization of the electron should simultaneously lower the anisotropies in both the g and A values. Our hyperfine equations can be further modified using molecular orbitals. We will not go into that much detail, although Maki and McGarvey did so in their original paper and most workers do prefer the more complete molecular orbital equations.

If one examines the spectrum of a powder or glass, the result is a broad absorption. The range of such absorptions is given by Eq. (8.16), but the intensity must be modified by the fact that the probability of an angle θ is proportional to $\sin\theta$ (see Fig. 1.4). In the derivative method one detects a signal that is proportional to the change in absorption with magnetic field. As a result, the important features in the normal derivative spectrum come from the extreme values of $H_0(\theta)$. One can see from Eq. (8.16) that extreme values obviously occur when $\theta = 0$ and $\theta = 90°$. These extreme values are clearly related to $g_{\|}$, $g_{\perp}$, $A_{\|}$, and $A_{\perp}$. Figure 8.11 shows the theoretical absorption and observed derivative

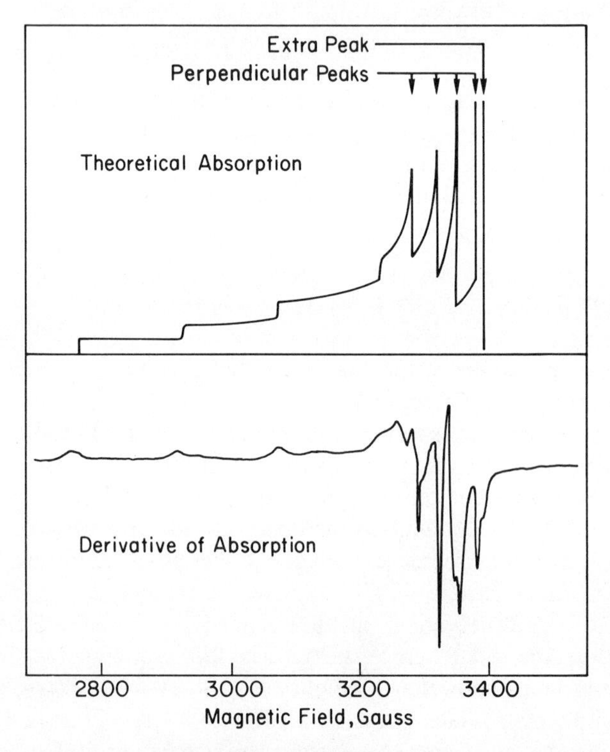

Figure 8.11. *Theoretical absorption spectrum and experimental first derivative, for copper acetylacetonate in a glass at* 138°K. (*From Gersmann and Swalen as adapted by McGarvey.*)

spectrum for copper acetylacetonate in a low-temperature glass. The low-field breaks in the spectrum are due to $\theta = 0$ and give $g_{\|}$ and $A_{\|}$, while the high-field breaks give $g_{\perp}$ and $A_{\perp}$. The extra break at high field is due to an extra extreme value given by Eq. (8.16), which is between $\theta = 0$ and $\theta = 90°$. The result of the assignment of this glass spectrum is also given in Table 8.4. One can see quite satisfactory agreement between the two methods, particularly considering the uncertainty due to the possible distortions of the complex in the toluene–chloroform glass.

With computer programs the assignment of a glass spectrum is a trivial problem for a system that can be fit by an axial Hamiltonian. In those cases where the molecular symmetry is low the angle ϕ must be included in Eq. (8.16). In this case not only are three g values and three A values necessary, but all calculations are more difficult. Fortunately, in second-rank tensors like g and A no more than three principal values are necessary to characterize them fully. One often finds that two of the principal values are close in value, so that $g_x \approx g_y$ in Eq. (4.1*a*) and $A_x \approx A_y$ in the complete Hamiltonian, including hyperfine interaction.

TRIPLET NAPHTHALENE

In Chapter 4 we discussed the fact that the first excited electronic states of many molecules are triplet states. In aromatic hydrocarbons these triplet states are sufficiently metastable so that EPR spectra can be obtained from them. The method for populating these triplet states is shown in Fig. 4.3. An intense light source at an allowed singlet absorption pumps a fraction of the molecules into the metastable triplet state. In order to have a long triplet lifetime it is best to dilute the aromatic in a glass or host crystal. The host can possess a triplet state, but it must have a higher energy so that the triplet excitation is trapped at the aromatic sites.

The dominant term in the spin Hamiltonian of a triplet state is the spin–spin or D term. In aromatic hydrocarbons this term is mainly due to the direct dipole–dipole interaction of the two unpaired electrons. Table 4.2 shows that in many aromatics $D/hc \approx 0.1$ cm^{-1}. Since this energy is close to the size of the common EPR quantum, such zero-field splittings are very important in the observed spectra. When a triplet state is placed in a strong magnetic field, it is split into three levels. The lowest level corresponds to $M_S = -1$, the middle level to $M_S = 0$, and the upper level to $M_S = +1$. Figure 4.4 shows such a splitting for H_0 both along x and z. The usual EPR selection rules are $\Delta M_S = \pm 1$. Figure 4.4 also shows the two $\Delta M_S = \pm 1$ transitions for each orientation with an assumed quantum corresponding to 0.30 cm^{-1}.

The H_0^0 value for $g = 2$ and a quantum of 0.30 cm^{-1} is 3200 gauss. One can see from Fig. 4.4 that for each orientation the pairs of lines are centered about this H_0^0 value. With the z-axis orientation of H_0 the two transitions are split exactly $2D$ apart, or in gauss units $2D/g\mu_B$ apart. With the x- or y-axis orientation just about one-half this splitting is obtained. Figure 8.12 shows an EPR spectrum for naphthalene in a host crystal of durene (1,2,4,5-tetramethylbenzene). In this crystal there are

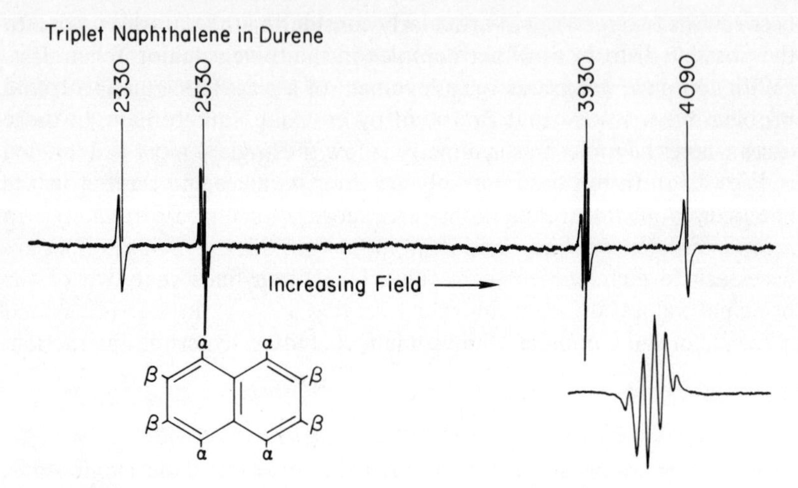

Figure 8.12. *Single-crystal spectrum for triplet-state naphthalene in durene at 77°K. (Courtesy of Varian Associates.)*

two possible sites for the naphthalene such that the two kinds of molecules are oriented so that the x axis of one kind is close to the z axis of the other. In Fig. 8.12 we show the EPR spectrum of triplet naphthalene with H_0 close to this common axis. The orientations are not perfect in Fig. 8.12, so that an accurate D value cannot be obtained from this figure. In addition, naphthalene has a small E value, which complicates the splitting pattern. The inner pair of lines in Fig. 8.12 can be seen to be a hyperfine multiplet. A five-line hyperfine multiplet, due mainly to the four α protons, is clearly resolved as shown in the expanded view in Fig. 8.12.

In Fig. 4.4 we also show an extra transition for the x,y orientation. This is from the bottom to the top level and it follows a $\Delta M_S = \pm 2$ selection rule. The only reason that this is possible is that when H_0 is not along z there are two possible axes for quantization—either space-fixed Z or molecule-fixed z. This means that M_S is not a good quantum number, or more accurately, that S_Z does not commute with H for H_0 not along z. The intensity of the $\Delta M_S = \pm 2$ transition is down by a factor of about $(D/g\mu_B H)^2$ from the normal $\Delta M_S = \pm 1$ transition. The importance of the $\Delta M_S = \pm 2$ transition is that it is nearly isotropic. In addition, it is close to being a "g" $= 4$ transition, or more accurately, a half-field transition since it is close to $H_0^0/2$.

When a triplet state is examined in a glass or a polycrystalline sample, the half-field transition is the strongest absorption since it is nearly isotropic. Under favorable conditions one can also detect, with derivative presentation, the breaks in the broad $\Delta M_S = \pm 1$ absorption. They occur when H_0 is along x, y, or z. The aromatics are favorable cases, for they have $10D \approx h\nu$ and the breaks are usually detectable. When $D \approx h\nu$, the half-field line becomes very broad and the $\Delta M_S = \pm 1$ absorption in a glass is so spread out that the breaks in the absorption are almost impossible to detect. When $D \approx 10h\nu$, even the single-crystal spectrum is hard to measure since few, if any transitions, are available in the energy-level diagram.

8.6. EPR Spectra in Gases

In Chapter 2 we saw that most atoms had electronic ground states possessing either spin or orbital moments. A number of atoms can be produced in the gas phase as a result of an electrical discharge. For these reasons it is particularly simple to study H, N, O, F, etc., as gases by EPR. At total pressures of less than 10 mm of mercury their absorptions are sharp, and accurate measurements can be made by EPR. The spin Hamiltonian for atoms can be written in accord with Chapter 2 as

$$\mathsf{H} = g_J \mu_B \mathbf{H}_0 \cdot \mathbf{J} + A\mathbf{I} \cdot \mathbf{J} - g_I \mu_N \mathbf{H}_0 \cdot \mathbf{I}. \tag{8.23}$$

When $L = 0$ and $J = \frac{1}{2}$, the A value is the result of Fermi contact interaction; when $L \neq 0$, however, the A value is primarily due to dipole–dipole interaction. The reason that dipole–dipole interaction can be simply written as a scalar product is that the electronic angular momenta process about **J**. As a result, the magnetic field at the nucleus must point along an axis defined by **J**. In some atoms A is as large as 1 cm^{-1}, and in these cases the Paschen–Back effect cannot be produced with ordinary magnetic fields.

The hydrogen atom is a good example. Figure 2.2 shows the energy levels for fields up to 3000 gauss and they are given for all fields in Eqs. (2.49) and (2.51). The normal EPR selection rules are $\Delta M_S = \pm 1$, $\Delta M_I = 0$. With these selection rules one obtains two transitions, which would center on H_0^0 and be A apart in the Paschen–Back limit. Since $A/g\mu_B$ is close to 500 gauss, the Paschen–Back effect is incomplete at fields of only 3000 gauss. For this reason the center of the H-atom spectrum is shifted toward lower field, but in this case the splitting is still equal to A even if the Paschen–Back effect is not complete. If H_1 is parallel to H_0, an additional transition is possible. It follows the selection rule $\Delta M_S = \pm 1$, $\Delta M_I = \mp 1$. This transition is lower in intensity by about $(A/2x)^2$. It has zero intensity in the Paschen–Back limit, but it can be of normal intensity at low magnetic fields.

DIATOMIC MOLECULES

A number of diatomic radicals have been studied by EPR. The kind of spectra that one observes depends a great deal on the nature of the electronic state of the radical. This is because a molecule has rotational angular momentum, and the coupling between the rotational angular momentum and the electronic angular momentum can be very important to the EPR spectrum. The most common electronic ground states for diatomic radicals are the $^2\Pi$ and $^3\Sigma$. Examples of $^2\Pi$ ground electronic states are NO, ClO, OH, SH, and BrO. Examples of $^3\Sigma$ ground electronic states are O_2, S_2, SO, SeO, and NF. The EPR spectra of almost all these radicals have been detected in the gas phase and, in some cases, they have quite complex spectra.

In Chapter 4 we discussed the magnetic properties of NO, and with this background it should be possible to understand its EPR spectrum. Figure 8.13 shows the EPR spectrum observed for NO gas at $P \approx 1$ mm of mercury and $\nu \approx 9$ GHz. Since all the rotational levels for NO shown in Fig. 4.7 are well populated at room temperature, it is surprising that all their Zeeman splittings give such a simple spectrum. In fact, the spectrum shown in Fig. 8.13 is a result of the Zeeman effect of only one of the rotational levels shown in Fig. 4.7. The other rotational levels have Zeeman splittings, at fields less than 20 kG, which are too small for transitions with $\nu \approx 9$ GHz. The center of the spectrum shown in Fig. 8.13 is close to 8 kG.

It is easy to understand why the rotational levels associated with the $^2\Pi_{1/2}$ electronic state have little or no Zeeman splittings. With Hund's case A, Eq. (4.11*a*) shows that $\mu_\parallel$ is almost zero for the $^2\Pi_{1/2}$ term. The first-order Zeeman effect is determined by the projection of $\mu_\parallel$ on the Z axis, and if $\mu_\parallel$ is zero we expect only very small Zeeman splittings. The $^2\Pi_{3/2}$ term, according to Eq. (4.11*a*), has a large value for $\mu_\parallel$, but we need to know its projection along the Z axis.

This calculation can be done using the vector model. We can write

$$\mathsf{H}(\text{Zeeman}) = -\boldsymbol{\mu} \cdot \mathbf{H}_0 = g_J \mu_B \mathbf{J} \cdot \mathbf{H}_0. \tag{8.24}$$

If we follow the vector-model diagram in Fig. 4.6*b*,

$$g_J = -\frac{\mu_z \cos(\boldsymbol{\Omega}, \mathbf{J})}{\mu_B \, |\mathbf{J}|}, \tag{8.25}$$

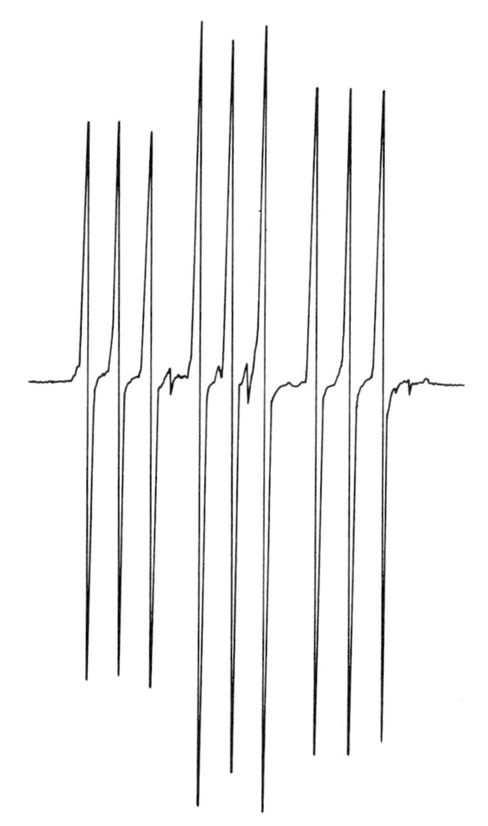

Figure 8.13. *Gas-phase spectrum of* NO *in its* $^2\Pi_{3/2}$ *state with* $J = \frac{3}{2}$. *The basic spectrum is three transitions with a* $\Delta M_J = \pm 1$ *selection rule. Each of these three transitions are split by* ^{14}N *hyperfine interaction.*

and since **N** is perpendicular to **Ω**,

$$\cos(\mathbf{\Omega}, \mathbf{J}) = \frac{\Omega}{|\mathbf{J}|}. \tag{8.26}$$

For NO, μ_z is identical to $\mu_{\parallel}$ given by Eq. (4.11*a*), and we obtain

$$g_J = \frac{(\Lambda + g_0\Sigma)\Omega}{J(J+1)}. \tag{8.27}$$

While this result assumes Hund's case A and it is also derived with the use of the vector model, it is remarkably accurate for NO. The $^2\Pi_{3/2}$ term has $\Lambda = 1$, $\Sigma = \frac{1}{2}$, and $\Omega = \frac{3}{2}$. Its lowest rotational level has $J = \frac{3}{2}$, as shown in Fig. 4.7. These values together with $g_0 = 2$ give $g_J = \frac{4}{5}$, and this agrees with an 8-kG resonance for $\nu \approx 9$ GHz. Since $J = \frac{3}{2}$ there are three possible transitions: $M_J = -\frac{3}{2} \rightarrow -\frac{1}{2}$, $M_J = -\frac{1}{2} \rightarrow \frac{1}{2}$, and $M_J = \frac{1}{2} \rightarrow \frac{3}{2}$.

Within the accuracy of Eq. (8.27) these three transitions would all have the same magnetic field for resonance, but this equation neglects the second-order Zeeman effect produced by $\mu_\perp$. The vector model becomes more difficult to utilize at this point, but more fundamental quantum-mechanical methods show that the Zeeman effect has second-order terms that depend upon M_J^2. These terms split the three $J = \frac{3}{2}$ transitions apart by about the 100 gauss, as shown in Fig. 8.13. The center transition belongs to the $M_J = -\frac{1}{2} \rightarrow \frac{1}{2}$ and it is expected to be more intense than the other two.

Hyperfine interaction with ^{14}N ($I = 1$) will also split each M_J transition into a 1:1:1 triplet. From Fig. 8.13 one can see that the hyperfine splitting at $\nu \approx 9$ GHz is smaller than the second-order Zeeman-effect splitting, as judged by the intensity pattern. One can also go back to Eq. (8.27) and see that for $J = \frac{5}{2}$, $g_J = \frac{12}{35}$. Its resonance field for $\nu \approx 9$ GHz should be close to 20 kG, which is above the range of most EPR apparatus.

Along the axis of the NO molecule there is an electric dipole moment in addition to the magnetic dipole moment that we used to obtain Eq. (8.27). With the selection rule $\Delta M_J = \pm 1$, both these dipoles are reoriented and can be used to induce these transitions. Detailed calculations show that the electric dipole moment is the more important one in intensity calculations, but the Zeeman effect is, of course, only produced by the magnetic dipole moment. At lower pressures some additional structure can also be detected in the NO spectrum, but we will not go into these details.

The OH radical has a more complicated EPR spectrum. Because of its faster rotation it is much closer to Hund's case B than is NO. This is a good example of where the vector model can fail, for only with more rigorous quantum mechanics can one deal with intermediate coupling. The $^3\Sigma$ molecules can have quite rich EPR spectra and O_2 has over 100 lines from 0 to 10 kG for $\nu \approx 9$ GHz. None of these lines is due to hyperfine structure since $I = 0$ for the major ^{16}O isotope.

Only a few polyatomic radicals have been studied by gas-phase EPR. They are difficult to produce in the gas phase and in some cases they have very weak EPR spectra. Although gas phase EPR can be used to observe some interesting and detailed spectra, it cannot be used for the wide variety of radicals that are observed in solids and in liquids.

References

A. Abragam and B. Bleaney, *Electron Paramagnetic Resonance of Transition Ions*, Oxford University Press, Inc., New York (1970). The major book on transition-metal ions in solids.

M. Bersohn and J. C. Baird, *An Introduction to Electron Spin Resonance*, W. A. Benjamin, Inc., Menlo Park, Calif. (1966). Elementary book which covers much of the field.

A. Carrington and A. D. McLachlan, *Introduction to Magnetic Resonance*, Harper & Row, Publishers, New York (1967). A good introduction to EPR with emphasis on clearly stated theory.

R. W. Fessenden and R. H. Schuler, "Electron Spin Resonance Studies of Transient Alkyl Radicals," *J. Chem. Phys.*, **39**, 2147 (1963). A long summary of alkyl radicals produced by irradiation of liquid hydrocarbons.

W. Low, *Paramagnetic Resonance in Solids*, Supp. 2 to *Solid State Physics* (Ehrenreich, Seitz and Turnbull, eds.), Academic Press, Inc., New York (1960). A rather detailed review of single-crystal work.

B. R. McGarvey, "Electron Spin Resonance of Transition-Metal Complexes," a chapter in *Transition Metal Chemistry*, Vol. 3 (R. L. Carlin, ed.) Marcel Dekker, Inc., New York (1966). An excellent review article.

G. E. Pake, *Paramagnetic Resonance*, W. A. Benjamin, Inc., Menlo Park, Calif. (1962). A general introduction. Particularly good for relaxation.

C. P. Poole, Jr., *Electron Spin Resonance*, John Wiley & Sons, Inc. (Interscience Division), New York (1967). A very complete account of experimental methods.

J. E. Wertz and J. R. Bolton, *Electron Spin Resonance*, McGraw-Hill Book Company, New York (1972). A particularly complete book with many examples of spectra.

Problems

8.1. Calculate the local field H^* that would produce an energy shift for an electron of 0.1 cm^{-1}. Relate this to a change in the Larmor frequency in radians per second. Calculate a line width in gauss using Eq. (6.25) with the assumption that H^* has a correlation time of 10^{-11} second.

8.2. Consider $Ni(H_2O)_6^{+2}$ in a crystal or in solution with octahedral coordination. Do you expect this ion to follow the Curie law? (Check your answer with Table 2.1 or in Ballhausen, page 142.) Do you expect an observable EPR spectrum in crystals at room temperature? Do you expect an observable EPR spectrum in water solutions? (Check your answer with Table 6.1.) Explain the difference between crystals and solution.

8.3.* Consider Co^{+2} in both octahedral and tetrahedral coordination. Which of these coordinations are more likely to give an observable EPR spectrum in a crystal at room temperature?

8.4. A common reference substance is peroxylamine disulphonate ion, $(SO_3)_2NO^{-2}$. It gives a sharp triplet hyperfine pattern in water solution with $A'(^{14}N) = 13.0$ gauss and $g = 2.0055$. What error in a g-value determination of H_0^0 of 3000 gauss would result if its g value were taken from the center of its triplet? What is the sign of this error? What would the error be if A' were 130 gauss instead of 13.0 gauss?

8.5. Use Eq. (8.8) to establish a power series for H_0(center) for $I = \frac{1}{2}$. Define H_0(center) as equal to the average H_0 value for $M_I = \pm\frac{1}{2}$. Confirm your result, which will not agree with $I = \frac{1}{2}$ in Eq. (8.9), by using the exact energy levels for $I = \frac{1}{2}$ as given by Eqs. (2.49) and (2.51).

8.6. Use VO^{+2} in water solutions as an example and calculate the actual H_0 values using Eq. (8.8) and $H_0^0 = 3000$ gauss. Which two of these H_0 values would you select to measure to determine the best A' value with the splitting formula based on the first-order Hamiltonian?

8.7. If a component of the H_1 field is parallel to H_0, the appropriate selection rule is $\Delta M_S = +1$, $\Delta M_I = -1$, and $\Delta M_S = -1$, $\Delta M_I = +1$. The intensity of these transitions is lower than the normal ones when $H_0^0 \gg A'$, but calculate their positions in the VO^{+2} spectrum.

8.8.* The radical anion of benzene has been observed in EPR and it was determined that $A'(H) = 3.75$ gauss. How many hyperfine lines should be observed and what intensity should they have? Also calculate a value for Q for the benzene radical anion.

8.9. Determine the best Q value for the tetracene radical cation if Hückel theory gives 0.034, 0.056, and 0.148 for the squares of the coefficients for the three carbon atoms responsible for the hyperfine structure.

8.10. The radical anion of naphthalene has two coupling constants, 1.83 and 4.90 gauss. Reconstruct the observed EPR spectrum. (*Note:* See Carrington and McLachlan, Chapter 8, if you want to check yourself.) The square of the Hückel coefficients for the α position (carbons 1, 4, 5, and 8) is 0.181; for the β position (carbons 2, 3, 6, and 7) it is 0.069. Use these to calculate a Q value for the naphthalene radical anion.

8.11. Calculate the A (solution) value in units of gauss for copper acetylacetonate.

8.12. On the basis of work with frozen glasses it has been determined that $g_{\|} = 2.400$ and $g_{\perp} = 2.099$ for $Cu(H_2O)_6^{+2}$. The absorption spectrum of Cu^{+2} in water gives a rather broad peak, which centers around 7000 Å. Use these data to calculate a λ value for Cu^{+2} in $Cu(H_2O)_6^{+2}$.

8.13. Make reference to Problem 8.1 and determine if it should be possible to observe the EPR spectrum of the triplet state of naphthalene in solution. This experiment was tried without success. [See S. I. Weissman, *J. Chem. Phys.*, **29**, 1189 (1958).]

8.14. The triplet-state energy levels with H_0 along x or y are given by $\epsilon = D/3$ and

$$\epsilon = \frac{-D}{6} \pm \sqrt{\left(\frac{D}{2}\right)^2 + g_x^2\mu_B^2H_0^2}.$$

Use this result to calculate the positions of the EPR signals in a glass spectrum of triplet naphthalene. These occur when H_0 is parallel to x, y, and z. Assume $g_x = g_y = 2.00$ and a constant frequency of 9.50 GHz.

8.15. How many EPR lines does one expect for the ClO radical in the gas phase? What are the approximate H_0^0 values for frequencies of 8.67 and 9.67 GHz? [See A. Carrington, P. N. Dyer and D. H. Levy, *J. Chem. Phys.*, **47,** 1756 (1967), Fig. 1.]

8.16. Use the vector model to predict the g_J value of EPR transitions of molecules in $^1\Delta$ states. This is a perfect example of Hund's case A. The lowest rotational state has $J = 2$; predict its g_J value with your formula. [See A. M. Falick, B. H. Mahan and R. J. Myers, *J. Chem. Phys.*, **42,** 1837 (1965).]

Solutions:

8.3. For octahedral coordination the electron configuration of Co^{+2} can be approximated as $(t_{2g})^5(e_g)^2$; this gives a $^4T_{1g}$ ground-state term. In a tetrahedral field one has $(e)^4(t_2)^3$, and this gives a $^4A_1$ ground-state term. Since the octahedral case has an orbital degeneracy of three, we expect fast relaxation at room temperature. The tetrahedral case has no orbital degeneracy and one should be able to observe its solid-state EPR near room temperature.

8.8. Six equivalent protons will give seven hyperfine lines with intensities that follow the binomial coefficients ($n = 6$) of 1:6:15:20:15:6:1. Since the two end lines are weak compared to the center, one often detects only five lines. With a spin density of $\frac{1}{6}$ on each carbon,

$$Q = 6 \times 3.75 = 22.5 \text{ gauss.}$$

Appendix

A.1. Important Physical Constants

The most recent recommended† values are

Electron mass (m_e) $= 9.1091 \times 10^{-28}$ g
Speed of light (c) $= 2.997925 \times 10^{10}$ cm-s^{-1}
Elementary charge (e) $= 4.80298 \times 10^{-10}$ esu
(e/c) $= 1.60210 \times 10^{-20}$ emu
Proton mass (m_p) $= 1.67252 \times 10^{-24}$ g

† NAS–NRC recommended values from the *National Bureau of Standards Technical News Bulletin*, **47,** 175 (1963). More precise constants are given by B. N. Taylor, W. H. Parker, and D. N. Langenberg, *Rev. Mod. Phys.*, **41,** 375 (1969).

Planck constant (h) = 6.6256×10^{-27} erg-s
Avogadro constant (N) = 6.02252×10^{23} mole^{-1}
Boltzmann constant (k) = 1.38054×10^{-16} erg-°K^{-1}
Bohr radius (a_0) = 5.29167×10^{-9} cm
Rydberg constant (R_∞) = 1.0973731×10^{5} cm^{-1}
Bohr magneton (μ_B) = 9.2732×10^{-21} erg-G^{-1}
Nuclear magneton (μ_N) = 5.0505×10^{-24} erg-G^{-1}
Electron moment correction $[(g_0/2) - 1] = 1.159615 \times 10^{-3}$
Proton Larmor constant ($\gamma/2\pi$) = 4.25770×10^{3} Hz-G^{-1}
(uncorrected for water protons) = 4.25759×10^{3} Hz-G^{-1}
Electron Larmor constant ($g_0\mu_B/h$) = 2.80246×10^{6} Hz-G^{-1}
(corrected to $g = 2$) = 2.79921×10^{6} Hz-G^{-1}

A.2. Units for Atomic Systems in Magnetic Fields

MAGNETIC UNITS

The conversion from mks to emu is complicated by the fact that mks is usually written in a rationalized form whereas emu (Gaussian) is not. A few comparison equations are

Nonrationalized (emu)		Rationalized (mks)	
$B = \mu_0(H + 4\pi M)$	or	$B = \mu_0(H + M)$	(A.1*a*)
$B = \mu_0(1 + 4\pi\chi)H$	or	$B = \mu_0(1 + \chi)H$	(A.1*b*)
$B = \mu_0 \dfrac{2iA}{r^3}$	or	$B = \dfrac{\mu_0}{4\pi}\dfrac{2iA}{r^3}$	(A.1*c*)

where the permeability of free space

$$\mu_0 = 1 \text{ (emu)} \qquad \text{or} \qquad \mu_0 = 4\pi \times 10^{-7} \text{ (mks)}. \tag{A.2}$$

The fundamental properties such as B and M have values that are equal in both a rationalized and nonrationalized mks, but H is decreased in value by 4π by rationalization. For this reason, mks results for B and M can be obtained from nonrationalized emu equations only if $\mu_0 = 10^{-7}$ is used, but the extra factor of 4π must be included whenever H is encountered. An important fact to remember is that χ in rationalized mks units is increased in value by 4π, owing to rationalization.

One of the best ways to handle the problem of units is to know the dimensions of the unit. The table below gives the units and dimensions of some important properties of magnetic systems. They are given in terms of mass, length, time, and μ_0. For mks we give the units in terms of the new international MKSA system. This system has been adopted as part of the Système International d'Unités or SI units. In this system B has the unit of tesla. It is equal to weber $\times$ (meter)$^{-2}$ in the old mks or 10^4 gauss in the emu. The gauss may still be used. The National Bureau of Standards has recommended† that it be retained as a special unit and has defined it as 10^{-4} T.

Property	Unit: emu	Unit: MKSA or SI	Dimensions: $[m]$	Dimensions: $[l]$	Dimensions: $[t]$	Dimensions: $[\mu_0]$
Mass	g	kg	1	0	0	0
Length	cm	m	0	1	0	0
Time	sec	sec (s)	0	0	1	0
Force	dyne	newton (N)	1	1	-2	0
Energy	erg	joule (J)	1	2	-2	0
Charge	abcoulomb	coulomb (A-s)	$\frac{1}{2}$	$\frac{1}{2}$	0	$-\frac{1}{2}$
Current	abampere	ampere (A)	$\frac{1}{2}$	$\frac{1}{2}$	-1	$-\frac{1}{2}$
B	gauss	tesla (T)	$\frac{1}{2}$	$-\frac{1}{2}$	-1	$\frac{1}{2}$
μ	erg/gauss	J/T	$\frac{1}{2}$	$2\frac{1}{2}$	-1	$-\frac{1}{2}$
H	oersted	A/m[a]	$\frac{1}{2}$	$-\frac{1}{2}$	-1	$-\frac{1}{2}$
χ_M	cm^3	m^3 [a]	0	3	0	0

[a] For these units rationalization has decreased the value of H by 4π and increased the value of χ_M by 4π.

With these dimensions one can compare relative values of emu and mks units:

$$\frac{\text{energy(mks)}}{\text{energy(emu)}} = \frac{m(\text{kg})}{m(\text{g})}\left[\frac{l(m)}{l(\text{cm})}\right]^2 = 10^{-7} \text{ or } 1 \text{ joule} = 10^7 \text{ ergs}.$$

When mks and emu units involve μ_0, a ratio of 10^{-7} should be used to eliminate rationalization:

$$\frac{B(\text{mks})}{B(\text{emu})} = \left[\frac{m(\text{kg})}{m(\text{g})}\right]^{1/2} \cdot \left[\frac{l(\text{m})}{l(\text{cm})}\right]^{-1/2} \cdot [10^{-7}]^{1/2} = 10^{-4}$$

or 1 tesla $= 10^4$ gauss. For a comparison of H and χ_M in the two systems a 4π must be introduced since rationalization has changed the

† National Bureau of Standards, "Policy for NBS Usage of SI Units," *J. Chem. Ed.*, **48**, 569 (1971).

size of the unit:

$$\frac{\chi_M(\text{mks})}{4\pi\chi_M(\text{emu})} = \left[\frac{l(\text{m})}{l(\text{cm})}\right]^3 = 10^{-6}$$

and

$$\chi_M(\text{mks}) = 4\pi \times 10^{-6}\chi_M(\text{emu}).$$

ENERGY

Many different units are used for energy in atomic systems. Each unit has a special advantage, depending upon the circumstance.

Atomic units. In theoretical calculations it is most convenient if the experimental values of the physical constants are never used. This can be accomplished by using atomic units. The atomic unit of length is a_0, the Bohr radius. The atomic unit for energy is equal to e^2/a_0, or twice the ionization energy of the hydrogen atom. However, some authors prefer to use the ionization energy of the hydrogen atom as their atomic unit for energy. In this case, it is equal to R_∞ and more properly called the Rydberg:

$$\begin{aligned}\text{atomic unit} &= e^2/a_0 = 2R_\infty \\ &= 4.35942 \times 10^{-11}\ \text{erg} \\ &= 27.2107\ \text{eV}\end{aligned}$$

Electron volts. The most common energy unit for the energy of the electrons in atoms is the electron volt. It is the energy associated with the transfer of the charge on one electron across a potential of one practical volt:

$$\begin{aligned}1\ \text{eV} &= 1.60210 \times 10^{-12}\ \text{erg} \\ &= 2.41804 \times 10^{14}\ \text{Hz} \\ &= 8.06573 \times 10^{3}\ \text{cm}^{-1}.\end{aligned}$$

Frequency. With Planck's equation $\epsilon = h\nu$ any energy may be correlated with a frequency. One cycle per second is now called a hertz and abbreviated Hz:

$$1\ \text{Hz} = h \cdot s^{-1} = 6.6256 \times 10^{-27}\ \text{erg}.$$

Wave number. In some branches of spectroscopy wavelengths are directly measured, and they can be corrected to that corresponding to a vacuum. Since the wavelength in vacuum λ follows $\lambda\nu = c$, wave number $\bar{\nu} = 1/\lambda = \nu/c$. In cgs the unit for wave number is cm^{-1} or reciprocal centimeter, but it can also be called a Kayser and abbreviated K:

$$1\ cm^{-1} = c \cdot cm^{-1} = 2.997925 \times 10^{10}\ Hz.$$

Magnetic field. In magnetic resonance, splittings are sometimes tabulated in gauss. With the relations $h\nu = g\mu_B H$ or $h\nu = \gamma_I H/2\pi$, magnetic field splittings can be converted into electron or nuclear frequency units:

$$1\ \text{gauss (water proton)} = 4.25759 \times 10^3\ Hz$$
$$1\ \text{gauss (electron, } g = 2) = 2.7992 \times 10^6\ Hz.$$

For zero-field splittings in EPR

$$1\ \text{gauss (electron)} = \frac{g}{2} \times 0.9337 \times 10^{-4}\ cm^{-1}.$$

A.3. Table of Nuclear Properties

The following information is taken from Varian Associates, **NMR Table,** 5th ed., Palo Alto, Calif. (1965). We have selected only the more important elements and isotopes. The complete table should be consulted for information about the other elements and isotopes. The listed NMR frequencies and magnetic moments may include a contribution from the chemical shift of a common form of the element.

Element	Isotope	% Natural abundance	NMR frequency (MHz/10^4 G)	μ_I	I
Aluminum	^{27}Al	100	11.094	3.6385	$\frac{5}{2}$
Beryllium	^{9}Be	100	5.983	−1.1773	$\frac{3}{2}$
Boron	^{10}B	18.83	4.575	1.8005	3
	^{11}B	81.17	13.660	2.6880	$\frac{3}{2}$
Bromine	^{79}Br	50.57	10.667	2.0991	$\frac{3}{2}$
	^{81}Br	49.43	11.499	2.2626	$\frac{3}{2}$
Carbon	^{13}C	1.108[a]	10.705	0.70220	$\frac{1}{2}$
Chlorine	^{35}Cl	75.4	4.172	0.82091	$\frac{3}{2}$
	^{37}Cl	24.6	3.472	0.68330	$\frac{3}{2}$
Chromium	^{53}Cr	9.54[a]	2.406	−0.47354	$\frac{3}{2}$
Cobalt	^{59}Co	100	10.103	4.6388	$\frac{7}{2}$
Copper	^{63}Cu	69.09	11.285	2.2206	$\frac{3}{2}$
	^{65}Cu	30.91	12.090	2.3790	$\frac{3}{2}$
Fluorine	^{19}F	100	40.055	2.6273	$\frac{1}{2}$

(continued overleaf)

Element	Isotope	% Natural abundance	NMR frequency (MHz/10⁴ G)	μ_I	I
Gold	^{197}Au	100	0.731	0.1439	$\frac{3}{2}$
Hydrogen	^{1}H	99.984	42.5759	2.79268	$\frac{1}{2}$
	^{2}H	0.016	6.53566	0.857386	1
Iron	^{57}Fe	2.245[a]	1.38	0.0903	$\frac{1}{2}$
Lead	^{207}Pb	21.11[a]	8.899	0.5837	$\frac{1}{2}$
Lithium	^{6}Li	7.43	6.265	0.82192	1
	^{7}Li	92.57	16.547	3.2560	$\frac{3}{2}$
Magnesium	^{25}Mg	10.05[a]	2.606	−0.85471	$\frac{5}{2}$
Manganese	^{55}Mn	100	10.553	3.4611	$\frac{5}{2}$
Mercury	^{199}Hg	16.86[a]	7.60	0.4979	$\frac{1}{2}$
	^{201}Hg	13.24[a]	2.80	−0.5513	$\frac{3}{2}$
Nickel	^{61}Ni	1.25[a]	3.79	0.746	$\frac{3}{2}$
Nitrogen	^{14}N	99.635	3.076	0.40358	1
	^{15}N	0.365	4.315	−0.28304	$\frac{1}{2}$
Oxygen	^{17}O	0.037[a]	5.772	−1.8930	$\frac{5}{2}$
Phosphorus	^{31}P	100	17.236	1.1305	$\frac{1}{2}$
Platinum	^{195}Pt	33.7[a]	9.153	0.6004	$\frac{1}{2}$
Potassium	^{39}K	93.08	1.987	0.39094	$\frac{3}{2}$
	^{41}K	6.91	1.092	0.21488	$\frac{3}{2}$
Scandium	^{45}Sc	100	10.344	4.7492	$\frac{7}{2}$
Silicon	^{29}Si	4.70[a]	8.458	−0.55477	$\frac{1}{2}$
Sodium	^{23}Na	100	11.262	2.2161	$\frac{3}{2}$
Sulfur	^{33}S	0.74[a]	3.266	0.64274	$\frac{3}{2}$
Thallium	^{203}Tl	29.52	24.33	1.5960	$\frac{1}{2}$
	^{205}Tl	70.48	24.57	1.6115	$\frac{1}{2}$
Titanium	^{47}Ti	7.75[a]	2.400	−0.78711	$\frac{5}{2}$
	^{49}Ti	5.51[a]	2.401	−1.1022	$\frac{7}{2}$
Vanadium	^{50}V	0.24	4.245	3.3413	6
	^{51}V	99.76	11.193	5.1392	$\frac{7}{2}$
Zinc	^{67}Zn	4.12[a]	2.664	0.87354	$\frac{5}{2}$

[a] The other abundant isotopes for these elements have $I = 0$.

A.4. Hamiltonian for Charged Particle Moving in Electric and Magnetic Fields

The first step in the derivation of the quantum-mechanical Hamiltonian operator is the formulation of the Hamiltonian function using the methods of classical mechanics. The Hamiltonian function $\mathscr{H}$ is defined by

$$\mathscr{H} = p_X v_X + p_Y v_Y + p_Z v_Z - \mathscr{L} \tag{A.3}$$

where p_i and v_i are the components of momentum and velocity, respectively.

The function $\mathscr{L}$ is called the Lagrangian and it is defined as

$$\mathscr{L} = \text{kinetic energy} - \text{potential energy}. \tag{A.4}$$

The momentum p_i and velocity v_i must be generalized conjugate coordinates and they must satisfy

$$p_i = \frac{\partial \mathscr{L}}{\partial v_i}\,. \tag{A.5}$$

ELECTRIC FIELD ONLY

In an electric field, such as provided by the Coulombic attraction of nuclei for electrons, there is an ordinary potential energy V. This potential energy depends only upon position and

$$\mathscr{L} = \tfrac{1}{2}m(v_X^2 + v_Y^2 + v_Z^2) - V, \tag{A.6a}$$

so that

$$p_i = \frac{\partial \mathscr{L}}{\partial v_i} = mv_i. \tag{A.6b}$$

If we combine the results of Eqs. (A.6*a*) and (A.6*b*) with Eq. (A.3), the result is

$$\mathscr{H} = \frac{p_X^2 + p_Y^2 + p_Z^2}{2m} + V. \tag{A.7}$$

In this form one can see that $\mathscr{H}$, in the absence of a magnetic field, is simply the sum of the kinetic and potential energies. The momentum in Eq. (A.7) is also the ordinary linear momentum. The quantum-mechanical Hamiltonian can be obtained by converting Eq. (A.7) into operator form and

$$\mathsf{H}(H = 0) = \frac{\mathsf{P}_X^2 + \mathsf{P}_Y^2 + \mathsf{P}_Z^2}{2m} + \mathsf{V}. \tag{A.8}$$

ADDITION OF MAGNETIC FIELD

A magnetic field will produce a force on a moving charge q. This force, however, is velocity dependent and its potential is not an ordinary scalar potential. The potential produced by a magnetic field is a vector potential **A** and it is defined by

$$\mathbf{A} = \text{curl}\,\mathbf{H} = \boldsymbol{\nabla} \times \mathbf{H}. \tag{A.9}$$

If we add this potential to the definition of $\mathscr{L}$ in Eq. (A.4), one finds that

$$\mathscr{L} = \tfrac{1}{2}m(v_X^2 + v_Y^2 + v_Z^2) - V + \frac{q}{c}(A_X v_X + A_Y v_Y + A_Z v_Z). \tag{A.10a}$$

The last term in this equation represents the velocity-dependent potential energy for a charge moving in a magnetic field. We must determine the generalized momentum and in this case

$$p_i = \frac{\partial \mathscr{L}}{\partial v_i} = mv_i + \frac{q}{c} A_i. \tag{A.10b}$$

This momentum is no longer equal to the ordinary linear momentum for we have an additional term that depends upon the magnetic field.

The substitution of Eqs. (A.10*a*) and (A.10*b*) into (A.3) gives the surprising result

$$\mathscr{H} = \tfrac{1}{2}m(v_X^2 + v_Y^2 + v_Z^2) + V. \tag{A.11}$$

This is equivalent to Eq. (A.7) and $\mathscr{H}$ is still only the sum of the kinetic and ordinary potential energies. The reason for this is that the vector potential caused by the magnetic field conserves energy. If, however, we want to use Eq. (A.11) to formulate the quantum-mechanical Hamiltonian operator we must take into account the difference between Eqs. (A.6*b*) and (A.10*b*). The reason for this is that the commutation rules of quantum mechanics are only valid for the proper set of generalized conjugate coordinates as defined by Eq. (A.5). As a result, we must formulate H in terms of the momenta p_i derived from Eq. (A.10*b*). The substitution of Eq. (A.10*b*) into Eq. (A.11) reveals that

$$\mathsf{H} = \frac{\mathsf{p}_X^2 + \mathsf{p}_Y^2 + \mathsf{p}_Z^2}{2m} + \mathsf{V} + \mathsf{H}'(H) \tag{A.12}$$

where

$$\mathsf{H}'(H) = -\frac{q}{2mc}(\mathsf{p}_X\mathsf{A}_X + \mathsf{A}_X\mathsf{p}_X + \mathsf{p}_Y\mathsf{A}_Y + \mathsf{A}_Y\mathsf{p}_Y + \mathsf{p}_Z\mathsf{A}_Z + \mathsf{A}_Z\mathsf{p}_Z) + \frac{q^2}{2mc^2}(\mathsf{A}_X^2 + \mathsf{A}_Y^2 + \mathsf{A}_Z^2). \tag{A.13}$$

Since the first two terms in Eq. (A.12) look very much like Eq. (A.8), they form what we shall call the "field-independent" Hamiltonian H_0. The part that must contain the explicit field dependence is Eq. (A.13).

In order to use Eq. (A.13) we must relate the vector potential **A** to the magnetic field **H** by means of Eq. (A.9). The only difficulty is that there is no unique set of A_X, A_Y, and A_Z even if we fix H along the Z axis. The choice of a set of components of the vector potential for a given H is known as the choice of gauge. Any gauge may be selected as long as it satisfies Eq. (A.9). The common gauge, as chosen by Van Vleck, is particularly convenient for atoms, and it is that

$$\mathsf{A}_X = -\frac{\mathsf{Y}H}{2} \qquad \mathsf{A}_Y = \frac{\mathsf{X}H}{2} \qquad \mathsf{A}_Z = 0. \tag{A.14}$$

With this gauge one obtains

$$\begin{aligned}\mathsf{H}'(H) &= -\frac{Hq}{2mc}(\mathsf{X}\mathsf{p}_Y - \mathsf{Y}\mathsf{p}_X) \\ &\quad + \frac{H^2q^2}{8mc^2}(\mathsf{X}^2 + \mathsf{Y}^2).\end{aligned} \tag{A.15}$$

One notes that the order of the products $\mathsf{p}_i\mathsf{A}_i$ and $\mathsf{A}_i\mathsf{p}_i$ were preserved in Eq. (A.13) since p_i and A_i may not always commute. In Eq. (A.15) this is not necessary since X commutes with p_Y and Y with p_X. The first term in Eq. (A.15) is called the paramagnetic contribution and the second is called the diamagnetic one. Because the gauge in Eq. (A.14) was arbitrary, this division between paramagnetic and diamagnetic is also arbitrary.

The paramagnetic term can be simplified, for orbital angular momentum **P** is defined by

$$\mathbf{P} = \mathbf{r} \times \mathbf{p}, \tag{A.16a}$$

so that

$$\mathsf{P}_Z = \mathsf{X}\mathsf{p}_Y - \mathsf{Y}\mathsf{p}_X. \tag{A.16b}$$

If we also use Eqs. (1.6) and (1.7), the first term in Eq. (A.15) is equal to

$$\mathsf{H}'(\text{para}) = \mu_B H \mathsf{L}_Z \tag{A.17}$$

where $q = -e$ for an electron. This result is identical to the energy of a magnetic dipole moment of value $-\mu_B\mathbf{L}$ in a field H along the Z axis. The angular momentum **L** is a generalized orbital angular momentum, for it is related to **p** by Eq. (A.16*b*), and **p** in Eq. (A.10*b*) is not the ordinary linear momentum.

The second term in Eq. (A.15) is called the diamagnetic one. It cannot be simplified further, so that for an electron

$$\mathsf{H}'(\text{dia}) = \frac{He^2}{8mc^2}(\mathsf{X}^2 + \mathsf{Y}^2). \tag{A.18}$$

The only restriction on this equation is that the center of the coordinate system which is used to define X and Y must be the same for $\mathsf{H}'(\text{dia})$ and $\mathsf{H}'(\text{para})$.

References

H. Goldstein, *Classical Mechanics*, Chaps. 1 and 7, Addison-Wesley Publishing Company, Inc., Reading, Mass. (1950).

J. H. Van Vleck, *The Theory of Electric and Magnetic Susceptibilities*, Chap. I, Oxford University Press, Inc., New York (1932).

A.5. Angular Momentum Matrix Elements and Selection Rules for Magnetic Resonance

The commuting operators define a set of eigenfunctions, and these eigenfunctions can be used to evaluate the matrix elements for angular momentum. One of the possible uses for such matrix elements is to determine the spectroscopic selection rules and the intensities of transitions in magnetic resonance.

These transitions are brought about by the interaction of an ac magnetic field, called H_1, and a magnetic dipole moment. In NMR the magnetic dipole moment is that of a nucleus and in EPR it is that of an electron. The usual circumstance is to orient H_1 in the XY plane so that the interacting moment is either μ_X or μ_Y. If we take H_1 as along X, the intensity of a transition from the state ψ to the state ψ' is proportional to

$$|\langle\psi|\,\mu_X\,|\psi'\rangle|^2; \tag{A.19a}$$

for NMR it is proportional to

$$\gamma^2(\text{nucleus})\,|\langle\psi|\,\mathsf{I}_X\,|\psi'\rangle|^2 \tag{A.19b}$$

or for EPR

$$\gamma^2(\text{electron})\,|\langle\psi|\,\mathsf{S}_X\,|\psi'\rangle|^2. \tag{A.19c}$$

The selection rules, in either case, are determined by the magnitude of the matrix elements required by these equations.

The commutation rules for angular momentum are also satisfied by the matrix elements for the angular momentum operators, so that the non-commutative algebra of operators can be replaced by matrix multiplication. Starting with Eq. (2.23) and matrix multiplication one has from the second commutation rule

$$i\hbar\langle\psi|\, \mathsf{P}_X\, |\psi'\rangle = \sum_{\psi} [\langle\psi|\, \mathsf{P}_Y\, |\psi''\rangle \langle\psi''|\, \mathsf{P}_Z\, |\psi'\rangle - \langle\psi|\, \mathsf{P}_Z\, |\psi''\rangle \langle\psi''|\, \mathsf{P} \quad |\psi'\rangle]. \tag{A.20}$$

If we choose wave functions that are eigenfunctions of P_Z,

$$\mathsf{P}_Z\psi = \hbar M\psi, \qquad \mathsf{P}_Z\psi' = \hbar M'\psi', \text{ etc.} \tag{A.2 }$$

The matrix elements of P_Z are then very simple and

$$\begin{aligned} \langle\psi|\, \mathsf{P}_Z\, |\psi\rangle &= \hbar M \\ \langle\psi'|\, \mathsf{P}_Z\, |\psi'\rangle &= \hbar M' \\ \langle\psi|\, \mathsf{P}_Z\, |\psi'\rangle &= 0. \end{aligned} \tag{A.22}$$

These results reduce the summations in Eq. (A.20) to one term each, and

$$i\langle\psi|\, \mathsf{P}_X\, |\psi'\rangle = (M' - M)\langle\psi|\, \mathsf{P}_Y\, |\psi'\rangle. \tag{A.23a}$$

If we apply this same technique to the third commutation rule in Eq. (2.23), one obtains

$$i\langle\psi|\, \mathsf{P}_Y\, |\psi'\rangle = (M - M')\langle\psi|\, \mathsf{P}_X\, |\psi'\rangle. \tag{A.23b}$$

Combining these two equations one sees that

$$\langle\psi|\, \mathsf{P}_X\, |\psi'\rangle = 0$$

unless

$$M' = M \pm 1. \tag{A.24}$$

The result of Eq. (A.24) for NMR gives the selection rule

$$\Delta M_I = M_I' - M_I = \pm 1 \tag{A.25a}$$

and for EPR

$$\Delta M_S = M'_S - M_S = \pm 1, \tag{A.25b}$$

or Eqs. (A.19*b*) and (A.19*c*) will otherwise give zero intensity. These selection rules would always be exact if the operators I_Z and S_Z commute with the Hamiltonian, for in this case the wave functions selected by Eq. (A.21) are also eigenfunctions of the Hamiltonian. Since H(Zeeman) for both electrons and nuclei commute with I_Z and S_Z, Eq. (1.13) is exact for noninteracting electrons and nuclei.

In the *AB* problems in Chapter 7, the selection rules $\Delta M_a = \pm 1$, $\Delta M_b = 0$, and $\Delta M_b = \pm 1$, $\Delta M_a = 0$ are exact if the Hamiltonian contains no spin–spin interaction. But since I_{Za} and I_{Zb} do not commute with $\mathbf{I}_a \cdot \mathbf{I}_b$, these rules are not exact when this interaction is included. We have seen that F_Z does commute with both $\mathbf{I}_a \cdot \mathbf{I}_b$ and H(Zeeman), and there should be a general selection rule for M_F. A calculation identical to our previous one shows that this selection rule is

$$\Delta M_F = \pm 1. \tag{A.26}$$

This is a very general selection rule and it can be applied to a number of problems in spectroscopy when the ac field is in the *XY* plane.

Detailed, calculations for the *AB* problem can give relative intensities of the four transitions for all ranges of J and x. These results depend upon the fact that

$$F_X = I_{Xa} + I_{Xb} \tag{A.27}$$

and the intensities are proportional to

$$|\langle\Psi| F_X |\Psi'\rangle|^2 \tag{A.28}$$

where these wave functions must be eigenfunctions of the *AB* Hamiltonian. The matrix that results from Eq. (A.28) is, in tabular form,

M_F	-1	0	0	1
-1	0	$1 + \sin 2\theta$	$1 - \sin 2\theta$	0
0	$1 + \sin 2\theta$	0	0	$1 + \sin 2\theta$
0	$1 - \sin 2\theta$	0	0	$1 - \sin 2\theta$
1	0	$1 + \sin 2\theta$	$1 - \sin 2\theta$	0

where $\sin 2\theta = +J/(x^2 + J^2)^{1/2}$. A factor of $\frac{1}{2}$ has also been removed from all the elements. The rows and columns correspond to an order in decreasing energy. These results were used to evaluate the intensities shown in Fig. 7.3.

Index